Problem Solving with
ADA

WILEY SERIES IN COMPUTING

Consulting Editor

Professor D. W. Barron

*Department of Mathematics, Southampton University,
Southampton, England*

Countess Ada Lovelace

Problem Solving with ADA

Brian Mayoh

*Computer Science Department,
Aarhus University, Denmark*

JOHN WILEY & SONS

Chichester · New York · Brisbane · Toronto · Singapore

Library of Congress Cataloging in Publication Data

Mayoh, B. H.
    Problem solving with ADA.

    (Wiley series in computing)
    Includes index.
    1. Ada (Computer program language) 2. Structured
programming. I. Title. II. Series
QA76.8.A15M38   001.64'24   81-14675
ISBN 0 471 10025 0   AACR2

British Library Cataloguing in Publication Data

Mayoh, Brian
    Problem solving with ADA—(Wiley series in
    computing)
    1. Ada (Computer program language)
    I. Title
    001.64'24   QA76.73.A35

ISBN 0 471 10025 0

Typeset by Activity, Salisbury, Wilts, and printed by
Pitman Press Ltd., Bath, Avon.

Contents

Preface

Habits are hard to change. For those learning to program a computer, an unfortunate choice of textbook or programming language may give an approach to solving problems that is a severe handicap in one's later career. For those already plagued by bad habits, the only hope is to attend courses or read books about the conceptual developments in the art of programming. This magic art is changing and the next generation of languages and computers will reflect these changes.

The structured programming approach to the art of solving problems on a computer has become very popular. We formulate this approach as: repeatedly split a problem into subproblems until the subproblems are easy to solve, then write the programs for solving the subproblems and combine these programs into a solution of the original problem. A textbook can only use this approach successfully if it uses a programming language which makes it easy to construct a large program from a multitude of small programs. The programming laguage ADA, which we use in this book, is just such a language. It is the product of development work by groups all over the world in the last five years. This work was financed by the US Department of Defense, and they are in a position to persuade almost all computer manufacturers to write ADA translators for their products. It is likely that ADA will be the language of most programs written in five years time; it is fortunate that ADA is also suitable for an introductory textbook.

In chapter 1 we look at how one can specify problems precisely and we introduce the divide-and-conquer approach to problem solving. Chapter 2 is primarily a description of the notions of algorithm, variables and parameters, but it also tells you how to convert algorithms into ADA programs and run them on a computer. In chapter 3 you will meet several powerful ways of combining solutions of small problems into solutions of large problems: choice, repetition, recursion, exceptions and parallelism. Chapter 4 is devoted to the slogan 'careful design of environments is the key to solving large problems'. You will meet environments for drawing pictures, editing texts, and using data bases. Chapter 5 explains the ADA type mechanism for finding conceptual errors in problem solutions. In chapter 6 you will meet various useful ways of structuring data: files, arrays, records and access structures. Chapter 7 illustrates the ADA concept of generic problem solutions by treating viii

the important practical problem of sorting and searching. Chapter 8 is about computers and people—the history of the computer revolution, and a discussion of the very real dangers that accompany this revolution. This chapter does not depend on any of the other chapters in the book, but it has been placed last in the hope that its message will be remembered when much of the book has been forgotten.

The author would like to thank his colleagues and students at the University of Aarhus for their valuable comments on this book. He is also grateful for the assistance of the designers of ADA at CII Honeywell Bull in Paris where the programs in this book were tested. He is most appreciative of his three artists, Niels Frey, Anne Nielsen and Rita Ljungberg, and his enthusiastic assistant Karen Møller.

Chapter 1

What is a Problem?

'The purpose of computing is insight, not numbers'
C. Hamming

As a person you are continually faced with intellectual and practical problems; as a member of a community you are involved with political, economic and social problems; at your work you are beset by other problems. The purpose of this book is to show you how a computer can solve some of your problems; the purpose of this chapter is to indicate the kinds of problems that computers can solve.

1.1 People and machines as problem solvers

We solve problems by thinking, but there seems to be two distinct modes of thought. When the left half of our brain is dominant, we are 'scientific', when the right half is dominant, we are 'artistic'. When we are thinking scientifically, we focus on language, analysis, order, laws and tools; when we are thinking artistically, we focus on patterns, synthesis, imagination and intuition. Both ways of thinking help us solve problems and both should be cultivated.

Computers are good scientific thinkers, they can swiftly carry out long tedious routine tasks without making mistakes. When a way of finding the solution of a problem can be formulated precisely in an appropriate language, a computer can find the actual solution quicker and more reliably than a person. If we think of the difference between walking and flying to some far off place, we realize that computers allow us to solve some problems we could not otherwise solve because they are too complicated or too time consuming. On the other hand there are many problems whose solutions require an artistic, creative way of thinking. When we are ill, we do not want a computer to diagnose our illness and prescribe our treatment, we want a doctor.

What comes next?

The best solution for many problems is 'man-machine interaction'—the computer busies itself with the details of data manipulation and analysis, but it refers to a person when it needs a decision that requires human judgement. With this in mind let us look at some typical problems.

Problem 1 Storm warning

When should we evacuate?

Some areas of Denmark, including a town of 10000 inhabitants, lie below sea level. When there is a storm and the sea threatens to overrun the protecting dykes, the police have to decide whether these areas should be evacuated. They make their decision on the basis of information about water levels, wind strengths and the like at various times and places. Because there is so much information, it is clear that a computer could help the police make better decisions.

What degree of interaction between computer and police should there be in this problem? Should the computer just remember the information it is given, or should it use some mathematical or statistical model to predict the probability of flooding? The answer is not clear.

Problem 2 Schizophrenia

The diagnosis of mental diseases is notoriously unreliable. Some years ago the World Health Organization made a survey of 5000 patients in seven countries, and they discovered enormous variations from country to country and even hospital to hospital. Each patient in the survey was asked 200 questions and their answers were recorded, together with their diagnosis, treatment and medical history. It became clear that 'schizophrenia' is probably several diseases, so a number of psychiatrists used the computer to find patterns in the survey data. What degree of interaction between computer and psychiatrist should there be in this problem? Should the computer just remember the survey data, or should it use some clustering method to suggest new diseases? The answer is not clear.

Problem 3 Rafts in the Pacific

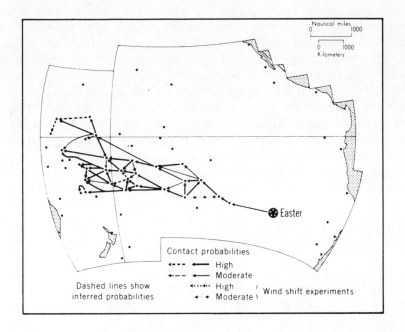

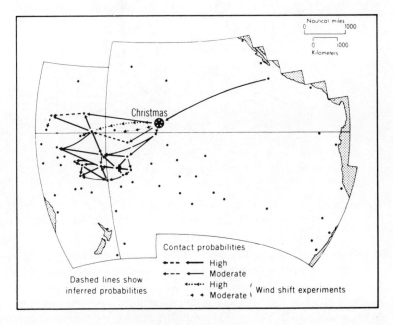

When do we sight land?

(From Michael Levison, R. Gerrard Ward, and John Webb, The Settlement of Polynesia: A Computer Simulation, University of Minnesota Press, Minneapolis. Copyright © 1973 by the University of Minnesota)

4

We do not know how the many widely separated islands of Polynesia came to be discovered and settled by a single people at a time when navigators of the 'civilised' world hardly dared to venture beyond the sight of land. As an argument for one theory, Thor Heyerdahl sailed from South America to Easter Island on the Kontiki raft. Others have used computers and other less dramatic arguments for their theories. One can isolate the factors that affect the course and survival of a small drifting boat; time and space changes in wind and current direction and speed; the location of reefs, islands and coasts, the course steered; the sailing qualities and seaworthiness of the vessel. Given this information and the starting position of a raft, the computer can predict where and when its passengers will sight land, if they do not run out of food or fall ill beforehand. In this problem solution the computer does not interact with the problem poser once she has specified the starting position of a raft.

Problem 4 Indus script

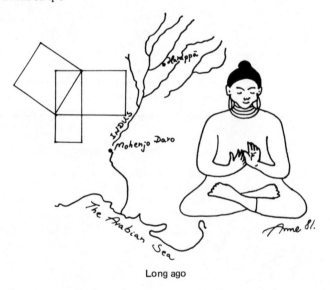

Long ago

Two thousand years ago there was a flourishing civilization in the Indus valley region. Today only the ruins of their enormous cities are left and the only remnants are a few hundred clay tablets with short inscriptions.

One can use a computer to remember these inscriptions and make simple comparisons, but human insight is needed to decipher the script. It so happened that the computer listed the common Indus word endings shortly before christmas one year, and the crucial human insight 'Indus script is based on the rebus principle' occurred during a vacation in Finland. The scholars showed the computer list to their professor and he recognised the symbols as an arrow, three yoke carriers, and a boat (the last because he had seen a drawing of a sumerian slave galley in one of his christmas presents). A Dravidian dictionary and the rebus principle gave 'to', 'of' and 'many' as the symbol meanings, so these symbols give case modifications of Indus words.

1.2 Problem specification

Before one can write a computer program to solve a problem, one has to have an exact specification of the problem to be solved. Usually the initial formulation of a problem is imprecise and the programmer has to make many decisions about unclearly formulated details before she really knows 'what the problem is'. As an example we can take the problem of making the drawings in this book. If you glance through the book, you will see that most of the illustrations are made by a computer but some are not. A more exact specification of our problem is:

P: Make the drawings in this book that can be constructed simply from straight lines, arcs and texts.

We need to be more exact than this. Must straight lines be solid or can they be dashed and dotted? What do we mean by 'constructed simply'. If we define a solid straight line or arc as a primitive segment, we can make our specification more precise by saying that we want computer programs that

— A: construct primitive segments
— B: combine several segments into one
— C: insert text in a segment
— D: scale, translate and rotate segments
— E: convert a segment into a drawing.

This specification is still not precise enough, but we will not go into finer detail now because we will return to the problem in later chapters. Instead we comment on the subproblem: Make the drawings in this book that consist of rectangles and oblongs connected by arrows. Because there are many such drawings, a computer solution of this subproblem would be very useful. Note that a solution of the original problem will most likely lead to a solution of the subproblem, but the converse is not true. Note also that the way you would describe a rectangle and oblong diagram to a friend is not the same as the way you would describe a general line and arc diagram. Because these description languages are so different, we will give a computer solution for each kind of diagram, and the author has in fact used two different computer programs for the diagrams in this book.

1.3 Divide and conquer

One effective method for making precise problem specifications is to split the problem into several subproblems. In the last section we split the problem

P: Make the drawings in this book that can be constructed simply from straight lines, arcs and texts

into the subproblems A, B, C, D, E. Before we solve these subproblems we should refine them further until we reach the point when

- the remaining subproblems have a solution which can be understood immediately;
- the way in which the solutions of subproblems can be combined into a solution of the original problem can be understood immediately.

A good guiding principle for what can be understood immediately is: our human minds can only grasp something directly, if it can be described clearly on a small piece of paper. Our decision to refine problem P into subproblems A, B, C, D, E would have been bad, if you could not immediately understand how a solution of the subproblems would give a solution of P.

If you do not use the divide and conquer methodology, your problem solutions may develop into an indigestible morass of computer programs that you cannot document adequately. When you want to use them again in six months' time, you may run into undiscovered errors that you cannot understand; when you want to modify them to solve a different but similar problem, you may get bogged down in obscure details. Occasionally there are reasons to abandon the strict divide and conquer methodology, but you should never forget the dangers of doing so.

Exercise

You are asked by your local authority to replan the bus routes in your neighbourhood. What information would you need in order to find the best solution of this problem? Specify precisely how the best solution of this problem could be found if the necessary information was available. Would you let a computer program find this best solution without interacting with those who have experience in planning bus routes? If not, how would you describe the interaction between planners and your program?

Exercise

Cryptography is a very important way of protecting information. Until 1976 all ways of encoding and decoding information required the sender and receiver to agree on a secret key K

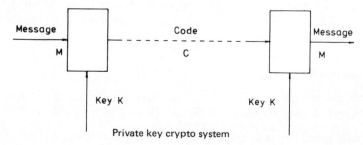

Solve the problem of encoding and decoding arbitrarily large messages on the assumption that you have a solution of the subproblem: encoding and decoding messages consisting of precisely 100 letters. In 1976 the first system appeared

which allowed anybody to send an encoded message by using the receiver's public key. The mathematically minded might enjoy finding the requirements on d, e, n for one public key solution for our subproblem

- C is the remainder of M^e when divided by n
- M is the remainder of C^d when divided by n

where e is the public key, d is the private key, M is the decoded message and C is the encoded message.

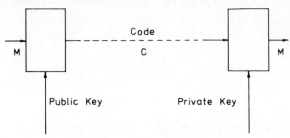

Public key crypto system

Chapter 2

Algorithms and Machines

In this chapter we present a way of solving problems on a computer

- devise an algorithm on the assumption that you already have various subalgorithms
- design the assumed subalgorithms
- convert the algorithms and subalgorithms into an ADA program.

Because computer systems vary greatly, you may have to read the manual for your local computer before you can feed it ADA programs.

2.1 What is an algorithm?

An algorithm is a precise description of how to solve a problem. If we have the problem:

(BISECTION)   find the midpoint of the line between two points A and B

then we can solve the problem with the algorithm

(Step 1) Draw circles with centers A and B.
(Step 2) Draw a line between the two points, where the two circles intersect. The midpoint of the line AB is the point where the two lines intersect.

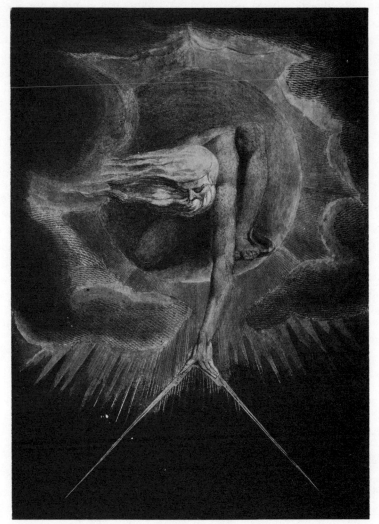

Blake's *'Europe'*

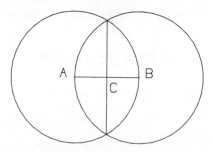

Bisection of line AB

This algorithm is suitable for people, because its two steps can be understood immediately by anybody who understands the original problem. The algorithm is not suitable for the computer, because it is not written in a language, that the computer understands, and it does not specify what should be done if the two circles do not intersect.

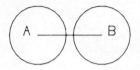

An unsuccessful attempt

In this situation you would draw larger circles, but this intuition is beyond the intelligence of the computer. Computers only understand simple, primitive algorithms but they are good problem solvers because programming languages allow one to combine primitive algorithms into complex, sophisticated algorithms. When you reach the end of this chapter, you will be able to write programs for solving complicated problems on the computer. The difficult and creative part of solving problems on a computer consists of refining algorithms into subalgorithms until the subalgorithms are so primitive that they can be expressed in a programming language. We can describe the refinement of an algorithm by a sequence of *structure diagrams*. The first structure diagram is a statement of the problem

BISECTION IS

──────────────────────►(FIND MIDPOINT OF LINE BETWEEN TWO POINTS)──────────────────►

Bisection structure diagram

and later structure diagrams are refinements of this:

BISECTION IS ALSO

──────────────►(STEP1)────────────────────────►(STEP2)──────────►

STEP1 IS

──────────────►(DRAW CIRCLES WITH CENTRES A AND B)──────────────►

STEP2 IS

──────────────►(DRAW LINE BETWEEN CIRCLE INTERSECTION POINTS)──────────►

Refinements of Bisection structure diagram

Intuitively the result of 'obeying the algorithm BISECTION' is:

− first obey the subalgorithm Step 1 to produce the drawing

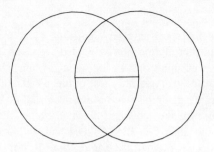

Bisection Step 1

— then obey the subalgorithm Step 2 to produce the drawing

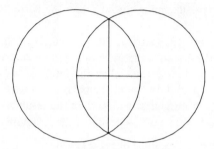

Bisection Step 2

This is imprecise because it does not describe what should be done if the sub-algorithm Step 1 produces the drawing

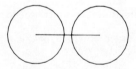

Bisection disaster

We can avoid this impasse by adding a new structure diagram to the sequence

BISECTION IS ALSO

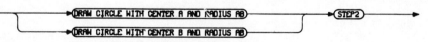

Better Bisection structure diagram

Not only does this structure diagram ensure that the subalgorithm Step 2 can always be obeyed, but it emphasizes the fact that either of the two circles can be drawn first.

One of the reasons for specifying an algorithm by a sequence of structure diagrams is that we can then define precisely what we mean by 'obeying an algorithm'. If we have a structure diagram for an algorithm, the names of various subalgorithms will be inside *oblongs* and these oblongs will be connected by *arrows* (for emphasis our structure diagrams will have arrows that do not connect oblongs but these have no effect on the meaning of the structure diagram). The second structure diagram for the algorithm BISECTION had one arrow connecting two oblongs, and these oblongs contained the names: Step 1, Step 2. Suppose we know the result of obeying the various subalgorithms in a structure diagram, when we know the arguments they are given. Then we can define the *possible results* of obeying an algorithm by:

- if an oblong has no incoming arrows, then the corresponding subalgorithm can be obeyed;
- if an oblong has incoming arrows, then the corresponding subalgorithm can be obeyed when there is an argument on each incoming arrow;
- if an oblong has outgoing arrows, then the result of obeying the corresponding subalgorithm is placed on one of the outgoing arrows;
- if an oblong has no outgoing arrows, then the result of obeying the corresponding subalgorithm is a possible result of the structure diagram.

These rules give precisely one result of obeying the algorithm

BISECTION IS ALSO

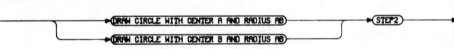

Better Bisection structure diagram

but they do not specify the order in which the subalgorithms are obeyed. One of the two possible orders is:

(1) obey the subalgorithm 'Draw Circle with center A and radius AB' and place the result

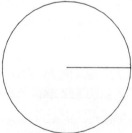

Better Bisection circle AB

on its arrow to the oblong named Step 2;

(2) obey the subalgorithm 'Draw Circle with center B and radius AB' and place the result

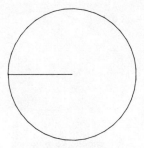

Better Bisection circle BA

on its arrow to the oblong named Step 2;

(3) collect the arguments on the two arrows to the Step 2 oblong, obey the subalgorithm Step 2 and deliver

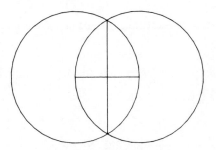

Better Bisection final result

as a possible result of the algorithm **BISECTION**.

Exercise

Devise an algorithm for drawing a line through a given point P that is parallel to a given line 1:

.... point P✳ ....

_____ Line L

Parallel line problem

What happens if you obey your algorithm when the point P is on the line 1?

2.2 Sequencing and parameters

Think of a tortoise sitting with a pen in its mouth on a large piece of paper:

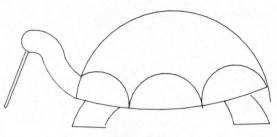

Symbolic Tortoise

Suppose we want to make the tortoise draw a square, and it knows how to obey the primitive algorithms:

Down   — the tortoise lowers its head, so its pen leaves a mark on the paper as it moves;

Up   — the tortoise raises its head, so its pen leaves no mark on the paper as it moves;

North, South, East, West
     — the tortoise makes one step in the named direction.

The problem for the tortoise is

POINT IS

Point structure diagram

and the tortoise will be able to solve the problem if it remembers the algorithm

POINT IS ALSO

——►(DOWN)———►(EAST)———►(NORTH)———►(WEST)———►(SOUTH)———►(UP)——►

Refined Point structure diagram

Let us return to reality. Computers are not tortoises but most of them can write characters on the screen of a display terminal. The computer knows where to write its character, because one position on the screen is distinguished from all others as the position of the cursor

A computer display

(Reproduced by permission of Lear Siegler, Inc.)

The computer knows how to obey the primitive algorithms

Down   — write marks as the cursor moves
Up   — write blanks as the cursor moves
North — move cursor one line up the screen
South — move cursor one line down the screen
East   — move cursor forward one position
West   — move cursor backward one position

so it can obey our algorithm Point and draw a square on the screen. Many computers also have a plotter

A computer plotter

(Reproduced by permission of scan-note A/S, Aarhus)

and they know how to obey the primitive algorithm

Down — lower plotter pen
Up — raise plotter pen
North — roll paper back one step
South — roll paper forward one step
East — move pen right one step
West — move pen left one step

so they can obey our algorithm Point and draw a square on their plotters. Unfortunately the square will probably be too small to be seen as a square, because plotters usually make steps in tiny fractions of an inch. We must extend the repertoire of primitive algorithms, if we want to make drawings as complicated as those in this book.

Once again imagine the computer as a tortoise with a pen in its mouth. Now suppose that it has a small brain in which it can remember a direction, and it can obey the primitive algorithms

Move(n) — the tortoise moves n steps in the remembered direction;
TurnTo(n) — the tortoise remembers direction n and forgets its previous direction;
Turn(n) — the tortoise adds n to its previous direction

where the letter *n* is a *parameter*. Whatever number we substitute for the parameter *n*, the tortoise can obey the three new primitive algorithms and it can interpret the number as a direction

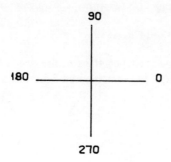

Directions as numbers

Suppose we want the tortoise to draw a large square. It can solve this problem

SQUARE IS

```
──────────────────────────►(DRAW LARGE SQUARE)──────────────────────────►
```

Square structure diagram

if it remembers the algorithm

SQUARE IS

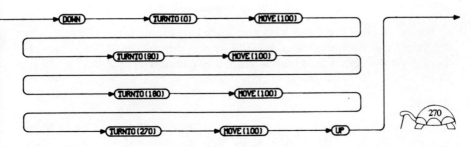

Refined Square structure diagram

For space reasons the direction the tortoise is remembering in its brain is drawn as a number in its stomach. Note that our algorithm would have given the same square as the algorithm Point if we had had 1 instead of 100 as the Move parameter.

Suppose we want our tortoise to draw a large triangle. It can solve the problem

TRIANGLE IS

```
──────────────────────►(DRAW LARGE TRIANGLE)──────────────────────────►
```

Triangle structure diagram

18

if it remembers the algorithm

TRIANGLE IS

Refined Triangle structure diagram

Notice that the drawing we get when the tortoise obeys this algorithm depends on the remembered direction; if the remembered direction is 0 and the algorithm is obeyed three times we get

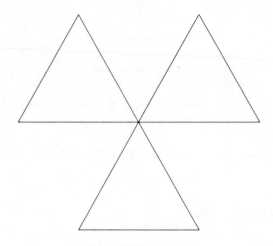

Thrice Triangle result

Notice also that this drawing has been made by a computer not a tortoise; if you look closely at the sloping lines, you will see that they are jagged. This is typical of the way plotters approximate curves which they cannot reproduce exactly.

Exercise

Give a detailed description of the way the tortoise draws the above diagram when it obeys the algorithm Triangle three times. Your description should include: then it obeys Move(100) to get

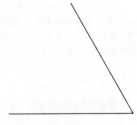

Partial Triangle result

Exercise

Devise algorithms for drawing

Problem drawings

Exercise

Devise an algorithm for a tortoise in a building that wants to get out of the only door:

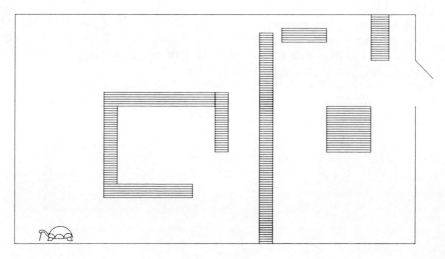

How to escape?

You can assume that the tortoise can recognize when it is at the door, and it knows how to obey the primitive algorithms:

RangeFinder(n)   — gives the distance from the tortoise's current position to the nearest obstacle in direction n;

DoorDirection   — gives the direction to the door from the tortoise's current position.

The tortoise should leave a trace of its way to the door when it obeys your algorithm

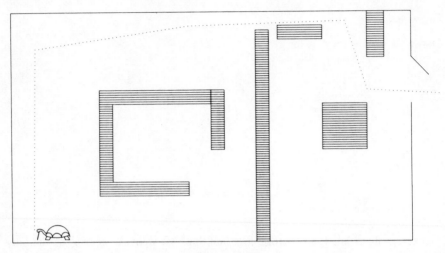

Escape route

A solution of this exercise is a computer simulation of animal behaviour. The author does not know of any algorithm which ensures that the tortoise always gets out of the building—with his algorithm the tortoise can get trapped in long blind passages (like eels and herrings in fishing nets).

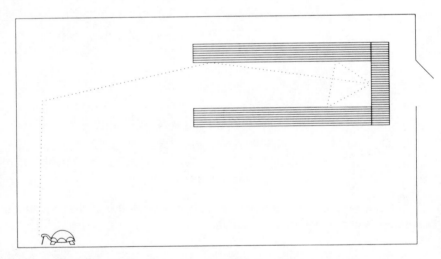

Trapped!

2.3 Subalgorithms

Suppose we want to solve the problem of drawing the corners of a square on the plotter

•   •

•   •

Corners of a square

Remembering the divide and conquer methodology of Chapter 1 we refine the problem into

CORNER IS

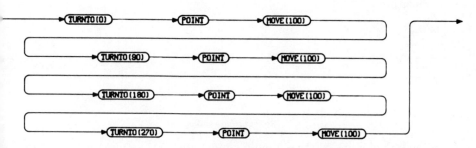

Corner structure diagram

and start thinking about the subalgorithm Point. Our first attempt might be

POINT IS NOT

Inadequate Point structure diagram

but we probably cannot see the point produced by the plotter version of this algorithm. If the computer obeys the algorithm

POINT IS ALSO

Adequate Point structure diagram

the resulting tiny square will look like a point and our problem is solved.

The divide and conquer methodology is an extremely powerful way of solving problems if we allow parameters in subalgorithms. Suppose we want an algorithm to draw the picture

Street of houses

Realizing that the houses are in the proportion 7:6:5:4:3:2:1 we refine the problem into:

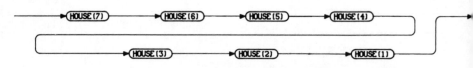

Street structure diagram

and start thinking about the subalgorithm House(*m*). Suppose we devise

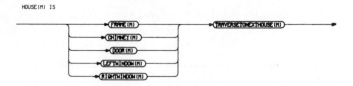

House structure diagram

Since five of the six algorithms in this structure diagram call for the drawing of a square, we should examine the sizes and positions of these squares, then make the refinement

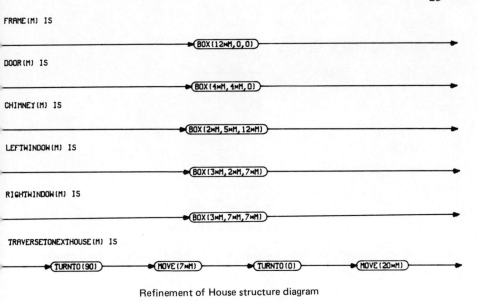

Refinement of House structure diagram

We complete the solution of the problem by defining Box(a,b,c) as a variant of last section's algorithm Square

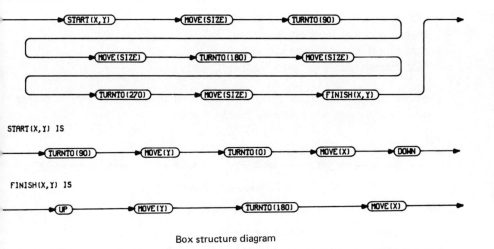

Box structure diagram

The subalgorithms Start and Finish ensure that the squares drawn by Box are positioned correctly.

Exercise

Devise an algorithm for drawing

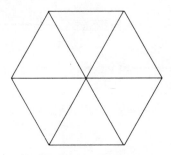

Another problem drawing

Does your algorithm draw the same lines twice? If it does, devise another algorithm that does not.

2.4 Values and variables

In this section we look more closely at the concept of a *value* and we explain how values can be kept in *variables* and used later. The algorithms we have seen so far remind one of the words by an ancient Persian poet and mathematician Omar Khayyam

'The Moving Finger writes and, having writ
Moves on, not all our Piety nor Wit
shall turn it back to cancel half a Line
Nor all your Tears wash out a Word of it'

Whenever our tortoise draws a line, the line is on the paper for ever. A description of what happens when we obey the algorithm

POINT IS ALSO

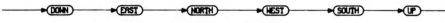

Point structure diagram

that is more precise than our earlier description is:

(1) obeying Down transmits the value *finished* to East;
(2) obeying East transmits the value *finished* to North and changes the result to

_____

East result

(3) obeying North transmits the value *finished* to West and changes the result to

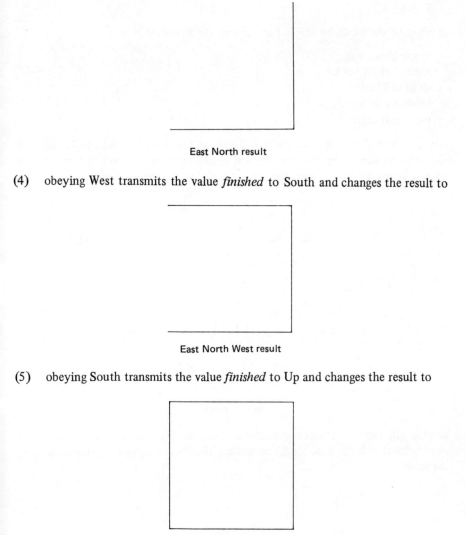

East North result

(4) obeying West transmits the value *finished* to South and changes the result to

East North West result

(5) obeying South transmits the value *finished* to Up and changes the result to

East North West South result

(6) obeying Up completes the algorithm.

The subalgorithms Down, East, North, West and South transmit a value to the next subalgorithm, but the fact that *finished* is the value transmitted is of no interest.

When we are divising an algorithm for solving a problem, we should distinguish between data and results of the solution and the values produced and used by subalgorithms. Imagine you are writing a letter. As you tap happily on the keys of a typewriter, it is obeying an algorithm that produces a sequence of characters as its result. This result might be the word on the next line

FENNEKE

it might be the sequence of 76 characters on the next five lines

tapping the type
writer keys SPACE
or LINEFEED
produces invisible
paper symbols;

it might be this sentence as a whole. Suppose you want to send the 76 character sequence to the computer and you are sitting at a display terminal, not a typewriter. Suppose your terminal like the author's has the key characters

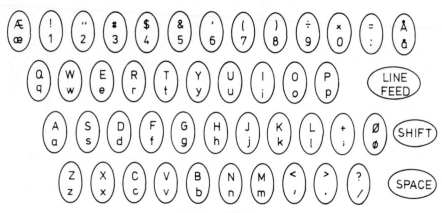

Key characters

so you can only communicate with the computer by sending sequences of these characters as values. You can communicate the 76 character sequence by sending the value

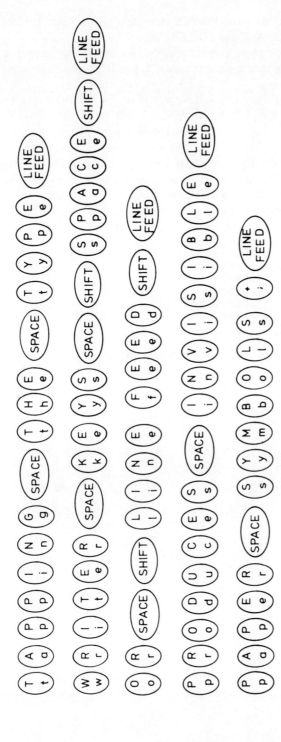

Key character sequence

28

Note that there are more than 76 characters in this value, because of the key characters SHIFT, SPACE and LINEFEED.

How can the computer communicate values to us? One answer is that it can draw values on a plotter. A cheaper answer is that it can place characters like

```
A  B  C  D  E  F  G  H  I  J  K  L  M  N  O  P  Q  R  S  T  U  V  W  X  Y  Z
a  b  c  d  e  f  g  h  ι  j  k  l  m  n  o  p  q  r  s  t  u  v  w  x  y  z
0  1  2  3  4  5  6  7  8  9     "  ♪  &  '  (  )  *  +  ,  -  .  /  :  ;  <  =  >  |
```

Display characters

on the screen of a display terminal. If we want a cheap, crude paper version of the results of obeying an algorithm, we can make the computer send characters to its 'lineprinter'; if we want a more elegant version of the results, we can make the computer send suitable values to some other 'device'. If your computer knows how to plot characters in different fonts, you may be able to give it the data

```
abcdefghijklmnopqrstuvwxyz
.normal
.serif
.gothic
.underline
.greek
.russian
```

Alphabet data

and get the result

abcdefghijklmnopqrstuvwxyz

abcdefghijklmnopqrstuvwxyz

abcdefghijklmnopqrstuvwxyz

abcdefghijklmnopqrstuvwxyz

αβηδεφγχιεκλμνοπθρστυφω ξ ψ ζ

абэдяфгхичклмнопшрстювщхуз

Alphabet result

The author's computer can send values to a photosetter with 118 characters in 8 fonts.

t	t	**t**	t	*t*	*t*	**t**	τ		M	M	**M**	M	*M*	*M*	**M**	M
o	o	**o**	o	*o*	*o*	**o**	o		L	L	**L**	L	*L*	*L*	**L**	Λ
h	h	**h**	h	*h*	*h*	**h**	η		R	R	**R**	R	*R*	*R*	**R**	P
n	n	**n**	n	*n*	*n*	**n**	ν		G	G	**G**	G	*G*	*G*	**G**	Γ
m	m	**m**	m	*m*	*m*	**m**	μ		I	I	**I**	I	*I*	*I*	**I**	I
l	l	**l**	l	*l*	*l*	**l**	λ		P	P	**P**	P	*P*	*P*	**P**	Π
r	r	**r**	r	*r*	*r*	**r**	ρ		C	C	**C**	C	*C*	*C*	**C**	Ǝ
g	g	**g**	g	*g*	*g*	**g**	γ		V	V	**V**	V	*V*	*V*	**V**	Θ
i	i	**i**	i	*i*	*i*	**i**	ι		E	E	**E**	E	*E*	*E*	**E**	E
p	p	**p**	p	*p*	*p*	**p**	π		Z	Z	**Z**	Z	*Z*	*Z*	**Z**	Z
c	c	**c**	c	*c*	*c*	**c**	ϑ		D	D	**D**	D	*D*	*D*	**D**	Δ
v	v	**v**	v	*v*	*v*	**v**	θ		B	B	**B**	B	*B*	*B*	**B**	B
e	e	**e**	e	*e*	*e*	**e**	ε		S	S	**S**	S	*S*	*S*	**S**	Σ
z	z	**z**	z	*z*	*z*	**z**	ζ		Y	Y	**Y**	Y	*Y*	*Y*	**Y**	Ξ
d	d	**d**	d	*d*	*d*	**d**	δ		F	F	**F**	F	*F*	*F*	**F**	Φ
b	b	**b**	b	*b*	*b*	**b**	β		X	X	**X**	X	*X*	*X*	**X**	X
s	s	**s**	s	*s*	*s*	**s**	∀		A	A	**A**	A	*A*	*A*	**A**	A
y	y	**y**	y	*y*	*y*	**y**	ξ		W	W	**W**	W	*W*	*W*	**W**	Ψ
f	f	**f**	f	*f*	*f*	**f**	φ		J	J	**J**	J	*J*	*J*	**J**	φ
x	x	**x**	x	*x*	*x*	**x**	χ		U	U	**U**	U	*U*	*U*	**U**	Υ
a	a	**a**	a	*a*	*a*	**a**	α		Q	Q	**Q**	Q	*Q*	*Q*	**Q**	Ω
w	w	**w**	w	*w*	*w*	**w**	ψ		K	K	**K**	K	*K*	*K*	**K**	K
j	j	**j**	j	*j*	*j*	**j**	σ		°	∅	£	—	*	◇	✳	⊃
u	u	**u**	u	*u*	*u*	**u**	υ		.	.	.	.	.	.		
q	q	**q**	q	*q*	*q*	**q**	ω		,	,	,	,	,	,	,	√
k	k	**k**	k	*k*	*k*	**k**	κ		0	0	**0**	0	*0*	*0*	**0**	⌐
T	T	**T**	T	*T*	*T*	**T**	T		1	1	**1**	1	*1*	*1*	**1**	𝒪
O	O	**O**	O	*O*	*O*	**O**	O		2	2	**2**	2	*2*	*2*	**2**	𝒫
H	H	**H**	H	*H*	*H*	**H**	H		3	3	**3**	3	*3*	*3*	**3**	𝒬
N	N	**N**	N	*N*	*N*	**N**	N		4	4	**4**	4	*4*	*4*	**4**	ℛ

Photosetter characters

If you have access to a photosetter, you can make programs that give results like

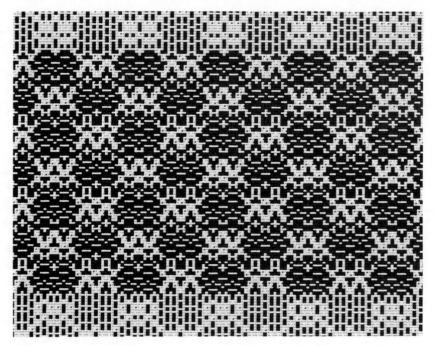

Pseudo carpet woven by one of the author's programs

Many devices can produce elaborate values from a small repertoire of characters; the above picture only uses four characters, reproductions of photographs in newspapers are no more than a host of tiny white and black dots. This whole book could have been produced on a computer with a 'matrix printer'—a printer that can place a dot or not place a dot at any given page position.

By now the reader should be convinced that there are many kinds of value and we need algorithms to convert from one kind of value to another. We would like an algorithm Normal to convert the key character sequence

Ⓐⓐ Ⓑⓑ Ⓒⓒ Ⓓⓓ Ⓔⓔ Ⓕⓕ Ⓖⓖ Ⓗⓗ Ⓘⓘ Ⓙⓙ Ⓚⓚ Ⓛⓛ Ⓜⓜ Ⓝⓝ Ⓞⓞ Ⓟⓟ Ⓠⓠ Ⓡⓡ Ⓢⓢ Ⓣⓣ Ⓤⓤ Ⓥⓥ Ⓦⓦ Ⓧⓧ Ⓨⓨ Ⓩⓩ

Key alphabet

to the value

abcdefghi jklmnopqrst uvwxyz

Display alphabet

we would like an algorithm Serif to convert our key character sequence to the value

abcdefghijklmnopqrstuvwxyz

Serif alphabet

we would like an algorithm Gothic to convert our key character sequence to the value

$$\mathfrak{a}\mathfrak{b}\mathfrak{c}\mathfrak{d}\mathfrak{e}\mathfrak{f}\mathfrak{g}\mathfrak{h}\,\mathfrak{i}\mathfrak{j}\,\mathfrak{k}\,\mathfrak{l}\,\mathfrak{m}\mathfrak{n}\mathfrak{o}\mathfrak{p}\mathfrak{q}\mathfrak{r}\mathfrak{s}\,\mathfrak{t}\,\mathfrak{u}\mathfrak{v}\mathfrak{w}\mathfrak{x}\mathfrak{y}\mathfrak{z}$$

Gothic alphabet

we would like an algorithm Underline to convert our key character sequence to the value

abcde f gh i j k l mnopqr s t uvwxyz

Underlined alphabet

we would like an algorithm Greek to convert our key character sequence to the value

$$\alpha\beta\eta\delta\varepsilon\varphi\gamma\chi\iota\epsilon\kappa\lambda\mu\nu o\pi\theta\rho\sigma\tau\upsilon\phi\omega\xi\psi\zeta$$

Greek alphabet

we would like an algorithm Russian to convert our key character sequence to the value

а б э д я ф г х и ч к л м н о п щ р с т ю в щ х у з

Russian alphabet

When we have all these algorithms we can reproduce the alphabet in many different fonts by obeying the algorithm

ALPHABET IS

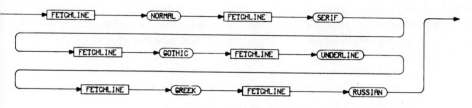

Alphabet structure diagram

We have used rectangles as well as oblongs in this structure diagram to emphasize the fact that the value produced by the subalgorithm FetchLine contains information that can be used by other subalgorithms. Throughout this book we will follow the convention—a rectangle is used in a structure diagram instead of an oblong, when *finished* is not the only possible result.

Let us turn to the idea of *storing a value in a variable.* Remember our tortoise pictures

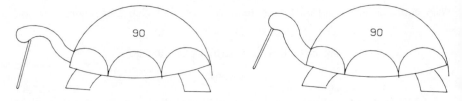

Tortoise keeping variables: direction and pen status

We can think of the number in the tortoise stomach as the last value stored in the variable 'direction'; we can think of the position of the tortoise head as the last value stored in the variable 'pen status'. The values of variables can be used by other subalgorithms if the tortoise can obey the primitive algorithms.

Direction   — produce the value last stored in the variable 'direction'
PenStatus   — produce the value last stored in the variable 'pen status'

and we change the definition of the possible results of obeying an algorithm to allow for structure diagrams with rectangles containing the names of variables. In our revised definition we also allow for oblongs like

```
TURNTO(90) IS ALSO
```

————————————————————————————▶(DIRECTION:=90)————————————————————————————▶

```
GETLINE IS ALSO
```

————————————————————————————▶(L:=NEWLINE)————————————————————————————▶

Structure diagrams: storing the values of variables

Such oblongs *assign* a value to a variable; the value given by obeying the subalgorithm after ':=' is stored in the variable whose name comes before ':='. With this in mind we reformulate the definition of the possible results of obeying an algorithm:

— if an oblong or rectangle has no incoming arrows, then the corresponding subalgorithm can be obeyed;
— if an oblong or rectangle has incoming arrows, then the corresponding subalgorithm can be obeyed when there is a value on each incoming arrow;
— if an oblong has outgoing arrows, then one of them acquires the value *finished* when the corresponding subalgorithm is obeyed;
— if a rectangle has outgoing arrows, then one of them acquires a value when the corresponding subalgorithm is obeyed; if the rectangle contains the name of a variable, then the value acquired by an arrow is the value last stored in the variable;
— when a subalgorithm is obeyed, the values of variables may change;
— when no subalgorithm can be obeyed, the value of the variable 'result' is a possible result of the algorithm.

We assume that every algorithm has variables 'data' and 'result', but we do not assume

- before any subalgorithm is obeyed, 'data' is the only variable that has a value;
- one can only change the value of 'data' by removing something;
- one can only change the value of 'result' by adding something;
- the initial value of 'data' determines the final value of 'result', so the order in which one obeys subalgorithms does not matter;

even although most algorithms satisfy these assumptions.

Variables allow one to manipulate values freely, but this freedom is not without its price. Algorithms for manipulating texts in computers have taken away the jobs of many secretaries and typographers. Many newspapers nowadays allow journalists to type an unjustified text like

```
Der kom en soldat marcherende henad
landevejen: Een, To. Een, To. Han
havde sit tornyster på ryggen og en
sabel ved siden, for han havde været i
krigen, og nu skulle han hjem. Så mødte
han en gammel heks på landevejen; hun
var så ækel, hendes underlæbe hang hende
lige ned på brystet. Hun sagde: God
aften soldat, hvor du har en pæn sabel
og et stort tornyster,du er en rigtig
soldat. Nu skal du få så mange penge,
du vil eje.
```

Unjustified text

which the computer transforms into a justified text like

```
Der  kom  en  soldat  marcherende  henad
landevejen:   Een,   To.   Een,   To.  Han
havde  sit  tornyster  på  ryggen  og  en
sabel  ved  siden, for han havde været i
krigen,  og nu skulle han hjem. Så mødte
han  en  gammel  heks på landevejen; hun
var så ækel, hendes underlæbe hang hende
lige  ned  på  brystet.  Hun  sagde: God
aften   soldat, hvor du har en pæn sabel
og  et  stort  tornyster,du er en rigtig
soldat.  Nu  skal  du få så mange penge,
du              vil              eje.
```

Justified text

Everyone, who writes a computer program, should be aware of the social problems the use of their program might provoke; the reader should not ignore the last chapter of this book.

Exercise

Redesign the algorithm

HOUSE(M) IS

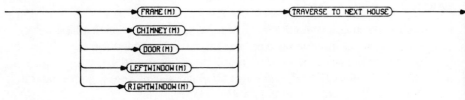

House problem

so that it uses variables. Your algorithm will still use the subalgorithms Frame, Chimney, Door, LeftWindow, RightWindow, TraverseToNextHouse.

2.5 Algorithms as programs

In this section we describe how algorithms can be converted into ADA programs, that can be 'run' on a computer. Remember the algorithm

POINT IS ALSO

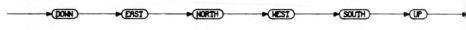

Point structure diagram

The first step in converting this algorithm into a program is to write the ADA *named unit*.

```
procedure Point is
begin
          Down;East;North;West;South;Up;
end       Point;
```

Point named unit

The first two lines and the last line show how an algorithm acquires an ADA name; the other line is given by putting semicolons for the arrows in the structure diagram. We can replace arrows by semicolons because none of the subalgorithms uses arrow values (there are no rectangles in the structure diagram). Since this is also true for the other algorithms in section 2.2 and 2.3, their ADA named units are:

```
procedure Square is
begin
          -- we can have comments
          Down; -- also here
          TurnTo(  0);Move(100);
          TurnTo( 90);Move(100);
          TurnTo(180);Move(100);
          TurnTo(270);Move(100);
          Up;
end       Square;

procedure Triangle is
begin
          Down;Move(100);Turn(120);Move(100);Turn(120);Move(100);Up;
end       Triangle;
```

```
procedure Corner is
begin
            TurnTo(   0);Point;Move(100);
            TurnTo(  90);Point;Move(100);
            TurnTo(180);Point;Move(100);
            TurnTo(270);Point;Move(100);
end         Corner;

 procedure Street is
begin
            House(7);House(6);House(5);House(4);House(3);House(2);House(1);
end         Street;


procedure House(m:INTEGER) is
begin
            Box(12*m,   0,    0);            -- Frame(m)
            Box( 2*m,5*m,12*m);             --Chimney(m)
            Box( 4*m,4*m,    0);            -- Door(m)
            Box( 3*m,2*m, 7*m);             --LeftWindow(m)
            Box( 3*m,7*m, 7*m);             --RightWindow(m)
        -- Now we traverse to the next house
            TurnTo(90);Move(7*m);TurnTo(0);Move(20*m);
end         House;

procedure Box(size,x,y:INTEGER) is
begin
        -- Start(x,y);
            TurnTo(90);Move(y);TurnTo(0);Move(x);Down;
        -- Make square of given size
            Move(size);TurnTo( 90);
            Move(size);TurnTo(180);
            Move(size);TurnTo(270);
            Move(size);
        -- Finish(x,y)
            Up;Move(y);TurnTo(180);Move(x);
end         Box;
```

Named units for Triangle, Square, Corner, Street, House and Box

In the named units for House and Box notice

— some subalgorithms appear as comments;
— the possible values of each parameter must be specified, when we write
':INTEGER' we specify that the value of the parameter must be a number;
— the arbitrary sequencing of the subalgorithms Frame, Chimney, Door, Left-
Window, RightWindow.

Understanding the structure diagram for an algorithm is easier than understanding its ADA named unit, but the judicious use of comments helps.

A named unit becomes an ADA program when it is supplied with an environment, which defines all the concepts the named unit uses but does not define. If we
have an environment Tortoise which defines the concepts Down, North, East,
South, West, Up—we can convert the named unit Point into the ADA program.

```
with        Tortoise
procedure Point is
begin
            Tortoise.Down  ;
            Tortoise.East  ;
            Tortoise.North;
            Tortoise.West  ;
            Tortoise.South;
            Tortoise.Up    ;
end         Point;
```

ADA Program: Point

The first line connects the program to an environment, and the prefix 'Tortoise.' tells where a concept is defined. We will put a frame around all the ADA programs in this book, as an encouragement to those who like to run programs on the computer and a challenge to those who know other programming languages. In our next ADA program we avoid prefix clutter by using ADA's *use* convention

```
with        Tortoise
procedure Square is
        use Tortoise;-- now concepts from the environment Tortoise
                   -- can be used without the prefix    Tortoise.
begin
            Down;
            TurnTo(  0);Move(100);
            TurnTo( 90);Move(100);
            TurnTo(180);Move(100);
            TurnTo(270);Move(100);
            Up;
end         Square;
```

ADA Program: Square

If you want to see how well your computer plotter can draw sloping lines, you could run the program

```
with        Tortoise
procedure Triangle is
        use Tortoise;
begin
            Down;Move(100);Turn(120);Move(100);Turn(120);Move(100);Up;
end         Triangle;
```

ADA Program: Triangle

Our next program shows how we can nest named units:

```
with       Tortoise
procedure Corner  is
      use Tortoise;
      -- because Corner uses the subalgorithm Point
      -- and Point is not defined in Tortoise
      -- we nest the named unit:
      procedure Point is
      begin
            Down;East;North;West;South;Up;
      end       Point;
   begin  -- Corner
         TurnTo(  0);Point;Move(100);
         TurnTo( 90);Point;Move(100);
         TurnTo(180);Point;Move(100);
         TurnTo(270);Point;Move(100);
   end       Corner;
```

ADA Program: Corner

The last of our programs has two nested named units:

```
with       Tortoise
procedure Street is
      use Tortoise;
      procedure Box(size,x,y:INTEGER) is
      begin
            TurnTo(90);Move(y);TurnTo(0);Move(x);Down;
            Move(size);TurnTo( 90);
            Move(size);TurnTo(180);
            Move(size);TurnTo(270);
            Move(size);
            Up;Move(y);TurnTo(180);Move(x);
      end       Box;
      procedure House(m:INTEGER) is
      begin
            Box(12*m,  0,   0);          -- Frame(m)
            Box( 2*m,5*m,12*m);          -- Chimney(m)
            Box( 4*m,4*m,   0);          -- Door(m)
            Box( 3*m,2*m, 7*m);          -- LeftWindow(m)
            Box( 3*m,7*m, 7*m);          -- RightWindow(m)
            -- now we traverse to the next house
            TurnTo(90);Move(7*m);TurnTo(0);Move(20*m);
      end       House;
   begin  -- Street
         House(7);House(6);House(5);House(4);House(3);House(2);House(1);
   end       Street;
```

ADA Program: Street

An ADA program is a sequence of display characters

```
A B C D E F G H I J K L M N O P Q R S T U V W X Y Z
a b c d e f g h ⌐ ⌐ j k L m n o p q r s t u v w x y z
0 1 2 3 4 5 6 7 8 9    "  ♪ & ' ( ) * + , - . / : ; < = > !
```

Display characters

but this sentence is a sequence of characters, that is not an ADA program. We need some way of distinguishing ADA programs from sequences of characters, that are not ADA programs. When you learn a natural language you meet grammatical rules like

a Noun followed by a Verb can be a Sentence

and you use these rules to group character sequences into categories like: Noun, Verb, Sentence. When you learn a programming language, you must expect syntax rules like:

a Number is a non-empty sequence of Digits
an Identifier is a letter followed by a sequence of Digits and Letters

which you can use to group character sequences into categories like Number and Identifier. We shall give syntax rules in the form of *syntax diagrams*

NUMBER IS

IDENTIFIER IS

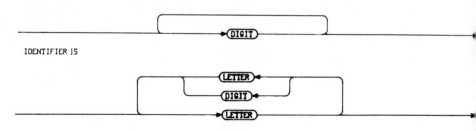

Number and Identifier syntax diagrams

The character sequence 'AB2' is an identifier, because this sequence corresponds to a path through the second syntax diagram; the character sequence '2AB' is not an identifier because it does not correspond to any path through the second syntax diagram. Syntax diagrams can contain rectangles as well as oblongs

HEAD IS

Head syntax diagram

A rectangle in a syntax diagram contains a character sequence that must be present, while an oblong indicates a requirement on character sequences that is defined by other syntax diagrams. Thus our syntax diagram gives

- 'procedure Box is' belongs to the category Head because 'Box belongs to the category Identifier;
- 'Box is' does not belong to the category Head because this character sequence does not start with procedure;
- 'procedure Box (size, x,y: INTEGER) is' does not yet belong to the category Head because identifiers cannot contain colons.

Since we want to use paramaters in ADA procedures, we need another syntax diagram

HEAD IS ALSO

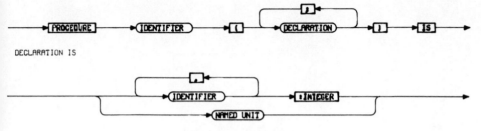

DECLARATION IS

Two more syntax diagrams

that puts '*procedure* Box (size, x,y: INTEGER) *is*' into the category Head. We close this section with a battery of syntax diagrams, which the reader can refer to when her computer refuses to run her programs because it has found a syntax error

NAMED UNIT IS

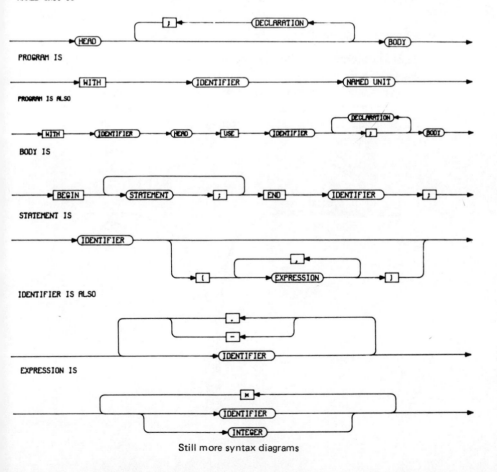

PROGRAM IS

PROGRAM IS ALSO

BODY IS

STATEMENT IS

IDENTIFIER IS ALSO

EXPRESSION IS

Still more syntax diagrams

Exercise

Devise ADA programs for drawing

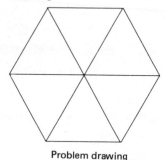

Problem drawing

Exercise

Check that the ADA programs in this section correspond to paths in the syntax diagram for the category Program.

Exercise

Construct syntax diagrams for

(a) trains which consist of a locomotive followed by one or more post-, goods- or passenger-sections where
— a goods-section consists of one or more goods-waggons
— a passenger-section consists of one or more passenger-waggons
— a post-section consists of a sorting-waggon, maybe followed by several letter-waggons.

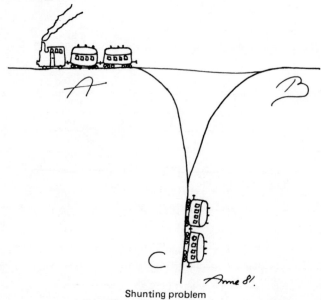

Shunting problem

(b) sequences of a's and b's that are possible descriptions of waggon shunting movements—'a' symbolizes a waggon movement from A to C, 'b' symbolizes a waggon movement from C to B, 'aabb' is possible, 'abba' is impossible.

(c) sequences of g's and b's in which there are equally many g's and b's—girls and boys at a formal dance.

2.6 Running programs on a computer

In this section we will describe how you can solve problems by running programs on your local computer. We will assume that your computer can understand and obey three primitive algorithms: DO, TRANSLATE, and EDIT. We assume you use

- DO to solve problems
- TRANSLATE to convert ADA programs to a form DO can understand
- EDIT to recover from your errors.

We have to make assumptions like these, because we cannot give the details of every reader's computer here; computers vary greatly and every computer has a manual the reader should acquire if she wants to run programs.

Before we describe how we can use the three-primitive algorithms

DO (data, result code)
TRANSLATE (data, result)
EDIT (data, result)

we should explain the significance of their parameters 'data' and 'result'. If we substitute 'screen' for 'result' and obey the primitive algorithm, then a new value will appear on the screen of our display terminal. This will also happen if a variable is substituted for 'result' *but* the computer will also assign the new value to the variable. If we substitute 'keyboard' for 'data' and obey the primitive algorithm, then it will use the value we type on the keyboard of our display terminal. If we substitute a variable for 'data' and obey the primitive algorithm, then it will use the value of the variable. If we substitute 'empty' for 'data' and obey the primitive algorithm, then it knows that there is no value it can use.

When the computer obeys the primitive algorithm

TRANSLATE (our program, our code)

it expects an ADA program as the value of the variable 'our program', it converts this program into a form DO can understand, and it assigns the converted program to the variable 'our code'. If the computer now obeys the primitive algorithm

DO (our data, our result, our code)

it looks at the values of the variables 'our data' and 'our code', and it assigns a value to the variable 'our result'. It could have obeyed

DO (our data, screen, our code)   —— if we did not want to keep the result
of running our program

DO (keyboard, our result, our code)   —— if we wanted to provide data from
the terminal

DO (empty, our result, our code)   —— if our program does not use data.

The optimist's way of running a program, that does not need any data, is to ask
the computer to obey

OPTIMIST IS

Optimist structure diagram

and hope that it translates the program the user types in at the terminal, then sends
the program results back to the terminal. Pessimists want to recover gracefully from
errors in their programs and data. They like the primitive algorithm

EDIT (data, result)

which takes the value of the parameter 'data', reads what the user types at her display terminal, and assigns the corrected value to the parameter 'result'. Pessimists
often ask the computer to obey the primitive algorithms

EDIT (empty, text)   —— so information from the display
terminal is assigned to the variable
'text'

EDIT (text, text)   —— so the value of the variable 'text'
is corrected using information from
the display terminal.

The pessimist's way of running a program, that does not need any data, is to ask
the computer to obey

PESSIMIST IS

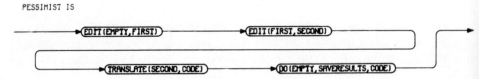

Pessimist structure diagram

If the computer finds an error in the pessimist's program, she can recover gracefully by asking the computer to obey

RECOVER IS

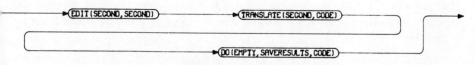

Recover structure diagram

When there are no more errors in her program, the pessimist will probably want to take her results home for careful analysis.

So far we have only explained how a computer can communicate with a user sitting at a display terminal; now we explain how a computer can present a 'hardcopy' value to a user, so she can take it home with her. We suppose that there is a variable for each output device on the computer, and substituting this variable for the parameter 'result' causes DO, TRANSLATE and EDIT to send a value to the output device. Thus we can ask the computer to obey the algorithm

EXTREME PESSIMIST IS

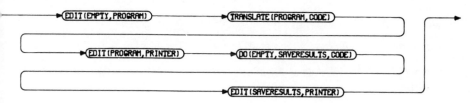

Extreme Pessimist structure diagram

and we expect hardcopy values of the variables 'program' and 'save results'.

Some output devices may produce nonsense, if they are sent inappropriate values; a step plotter may produce gibberish if it is sent display characters other than

DU   NS   EW

because it can interpret these characters as instructions to lower, raise or move its pen, but it cannot interpret other characters. You should always put a comment in your programs specifying the intended output device, as we have done in

```
with        Tortoise
procedure Triangle is
        use Tortoise;          -- intended results on plotter
        size:INTEGER;          -- we declare a variable
begin
        Get(size);             -- data gives value to variable
        Down;Move(size);Turn(120);Move(size);Turn(120);Move(size);Up;
end        Triangle;
```

Modified ADA program: Triangle

This program can draw triangles of many sizes because the value given to 'result' depends on the value supplied by 'data'. The program uses the ADA primitive Get which is available in all environments. There is another primitive which is always available

Put – for putting a number or character in 'result'

Sometimes it is convenient to combine successive calls of Put, to write Put("Todayis") instead of

Put("T"); Put("o"); Put("d"); Put("a"); Put("y"); Put(" "); Put("i"); Put("s");

Another useful convention allows us to divide 'result' into lines; we can write NewLine when we want to start a new line in 'result'.

Note for readers with access to Tortoise

You should run some of the programs in this chapter before reading on.

Note for readers who have access to ADA but not Tortoise

You should try to run a program on your computer before reading on. One possibility is

```
with       Text_IO
procedure SimultaneousEquations is
       use Text_IO; -- an environment known by all ADA systems
       -- the next line is necessary for obscure reasons
       -- we discuss in chapter 7
           package IO is new INTEGER_IO(INTEGER);use IO;
       a,b,c,aa,bb,cc,d:INTEGER;
begin
           Get(a);Get(b);Get(c);Get(aa);Get(bb);Get(cc);
           d := a*bb -aa*b;
           Put("The solution of the equations");          NewLine;
           Put( a);Put("x +");Put( b);Put("y =");Put( c);NewLine;
           Put(aa);Put("x +");Put(bb);Put("y =");Put(cc);NewLine;
           Put(" is x = ");Put((c*bb-cc*b)/d);            NewLine;
           Put("     y = ");Put((cc*a-c*aa)/d);           NewLine;
end        SimultaneousEquations;
```

ADA program: Simultaneous Equations

Note for readers who do not have access to ADA

You should try to convert some of our algorithms to programs in a language your computer understands, then run the programs.

Exercise

Our algorithms—Optimist, Pessimist, Extreme Pessimist—were designed for running programs that need no data. How should they be changed if we want to run a program that needs data?

Chapter 3

Flow of Control

All of the algorithms you have seen so far have been either line structured like

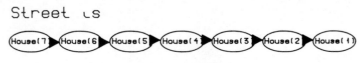

Line structured diagram

or tree structured like

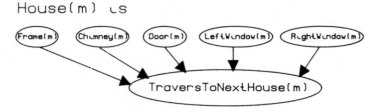

Tree structured diagram

In this chapter we will describe how to design algorithms with a more elaborate structure.

3.1 Choice

We often want an algorithm to use the value, generated by a subalgorithm, to choose between a number of possible next subalgorithms. The algorithm

Structured diagram with choice

48

uses the value given by the subalgorithm Generate Value to choose between Point, Square and Triangle. If the value is 1, the choice falls upon the subalgorithm Point; if the value is 4, the choice falls upon the subalgorithm Square; otherwise the choice falls upon the subalgorithm Triangle. In our next algorithm we introduce two more ways of prescribing values on arrows

IS

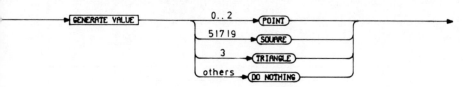

Another Structure diagram with choice

If the value generated is 3, then the subalgorithm Triangle is chosen, if the value generated lies between 0 and 2, then the subalgorithm Point is chosen; if the value generated is 5 or 7 or 9, the subalgorithm Square is chosen; otherwise the subalgorithm Do Nothing is chosen.

The results of our two algorithms are uninteresting but we will convert them to ADA programs even so. We begin by writing the ADA named units:

```
procedure A is
begin
        case GenerateValue() is
                when    1        =>      Point;
                when    4        =>      Square;
                when    others   =>      Triangle;
        end  case;
end     A;

procedure B is
begin
        case GenerateValue() is
                when    0..2     =>      Point;
                when    5|7|9    =>      Square;
                when    3        =>      Triangle;
                when    others   =>      DoNothing;
        end  case;
end     B;
```

Named units with choice

These named units become ADA programs when they are embedded in an environment Geometer, which defines Point, Square and Triangle

```
with        Geometer
procedure A is
        use Geometer; -- results on plotter
        n:INTEGER;
begin
        Get(n);
        case n is
                when  1     => Point;
                when  4     => Square;
                when others => Triangle;
            end case;
    end     A;
```

```
with        Geometer
procedure B is
        use Geometer; -- results on plotter
        n:INTEGER;
begin
        Get(n);
        case n is
                when  0..2  => Point;
                when 5|7|9  => Square;
                when  3     => Triangle;
                when others => null;
        -- null is ADA's algorithm for doing nothing
            end case;
    end     B;
```

ADA programs with number choice

You may prefer a variant of these programs that uses character instead of number values

```
with        Geometer
procedure A is
        use Geometer; -- results on plotter
        c:CHARACTER; -- note CHARACTER for INTEGER
begin
        Get(c);   -- reads CHARACTER value from data
        case c is
                when   'p' => Point;
                when   's' => Square;
                -- CHARACTER values in single quotes
                when others=> Triangle;
            end case;
    end     A;
```

```
with       Geometer
procedure B is
       use Geometer; -- results on plotter
       c:CHARACTER;
begin
       Get(c);
       case c is
           when  'p'..'r'   => Point; -- also 'q'
           when  's'|'v'|'c'=> Square;
           -- CHARACTER value 'c' differs from variable c
           when  't'        => Triangle;
           when  others     => null;
       end  case;
end        B;
```

ADA programs with character choice

Sometimes character values are more natural than number values; numbers would be misleading in the algorithm

PHOTOSETTER TEST

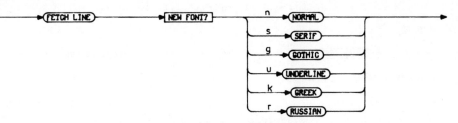

Photosetter Test structure diagram

and its named unit

```
procedure PhotosetterTest is
       L:LINE;
begin
       L := Dataline();
       case LineToCharacter() is
           when  'n'        =>  Normal;
           when  's'        =>  Serif;
           when  'g'        =>  Gothic;
           when  'k'        =>  Greek;
           when  'r'        =>  Russian;
           when  'u'        =>  Underline;
           when  others     =>  Put("unknown font");
       end  case;
end        PhotosetterTest;
```

Photosetter Test named unit

Sometimes it is natural to use values, that are neither numbers nor characters when choosing between possible next algorithms. In some computerized patient journal systems doctors and nurses have to remember number codes for diseases because the programmers were too lazy to use 'disease names' as values. Later we

52

shall describe the many **ADA** facilities for defining and manipulating values of various types; Now we close this section with syntax diagrams for the **ADA** choice mechanisms;

STATEMENT IS ALSO

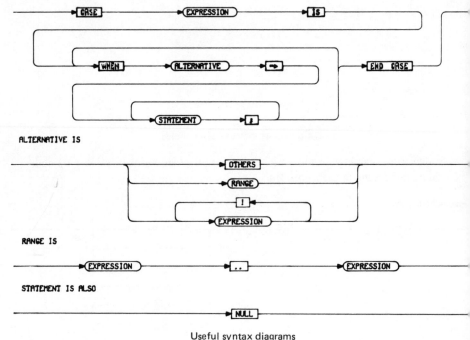

Useful syntax diagrams

Exercise

Devise an ADA program for drawing a square or triangle of arbitrary size. You can assume that the environment Geometer provides subalgorithms Square(size: INTEGER) and Triangle(size: INTEGER).

3.2 Repetition

Most useful algorithms contain subalgorithms that are used several times

LOOP IS

Repetitive structure diagram

If we want to solve the problem of drawing spirals like

53

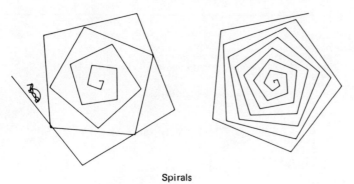

Spirals

we might design the algorithm

SPIRAL IS

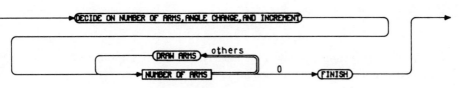

Spiral structure diagram

and write the named unit

```
procedure Spiral is
        number_of_arms,angle_change,increment,length:INTEGER;
begin
        Get(number_of_arms);Get(angle_change);Get(increment);
        length:=0;Down;
        loop
                case number_of_arms is
                        when    0      => exit;
                        when    others => DrawArm;
                                number_of_arms:=number_of_arms-1;
                end case;
        end loop;
        Up; -- the subalgorithm Finish
end     Spiral;
```

Spiral named unit

It is not difficult to write the named unit for the subalgorithm Draw Arm

```
procedure DrawArm is
begin
        length := length + increment;
        Move(length);
        Turn(angle_change);
end     DrawArm;
```

Draw Arm named unit

Combining these two named units gives an ADA program for drawing spirals

```
with        Tortoise
procedure Spiral is
        use Tortoise; -- results on plotter
        number_of_arms,angle_change,increment,length:INTEGER;
        procedure DrawArm is
        begin
                length := length + increment;
                Move(length);
        end       DrawArm;
begin --   Spiral
        Get(number_of_arms);Get(angle_change);Get(increment);
        length := 0;Down;
        loop
                case number_of_arms is
                        when    0     => exit;
                        when    others => DrawArm;
                                number_of_arms:=number_of_arms-1;
                end case;
        end   loop;
        Up;
end       Spiral;
```

ADA program: Spiral

In the last section you saw a program Photosetter Test, which would have been more useful if we had used repetition

```
with        BookProduction
procedure PhotosetterTest is
        use BookProduction; -- results on photosetter
        l:LINE;
begin
        loop
                l := DataLine();
                case LineToCharacter() is
                        when    'n'     => Normal;
                        when    's'     => Serif;
                        when    'g'     => Gothic;
                        when    'k'     => Greek;
                        when    'r'     => Russian;
                        when    'u'     => Underline;
                        when    others  => exit;
-- instead we could use 'e' for exit and give a helpful
-- message when we see a letter other than:e,n,s,g,k,r,u
                end   case;
        end   loop;
end       PhotosetterTest;
```

ADA program: Photosetter Test

One way of specifying the algorithm behind this program is to give a structure diagram, but a better way is to give the decision table

control character	'n'	's'	'g'	'k'	'r'	'u'	others
action	Normal repeat	Serif repeat	Gothic repeat	Greek repeat	Russian repeat	Underline repeat	

This decision table specifies the action to be taken for each possible control character. Decision tables can be used to specify more complicated algorithms if we allow more rows above and below the line. If we provide a row below the line for each subalgorithm, we can specify our algorithm by the decision table.

control character	'n'	's'	'g'	'k'	'r'	'u'	others
Normal	1						
Serif		1					
Gothic			1				
Greek				1			
Russian					1		
Underline						1	
repeat	2	2	2	2	2	2	

where the numbers below the line give the order in which subalgorithms should be obeyed. What can we gain by allowing more rows above the line in a decision table? One possibility is to allow a row for any subalgorithm that generates a value. A rather artificial example is

Generate First			1		4		others	
Generate Second	0..2			others	5!7!9!	others	3	others
Point	1							
Square					1			
Triangle							1	
Repeat	2			1	2		1	2

instead of the structure diagram

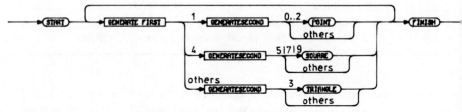

Structure diagram version of a decision table

When we are designing algorithms we often want to compare values generated by two subalgorithms; if the two values are the same we want to do one thing, if they are different we want to do another. Suppose we allow rows like

Value generated by subalgorithm A = Value generated by subalgorithm B	TRUE	FALSE

above the line in a decision table. Then we can specify the loop in our spiral drawing algorithm by the decision table

number_of_arms = 0	TRUE	FALSE
Draw arm		1
Repeat		2

Like most programming languages ADA allows one to use value comparisons to escape from a repeating subalgorithm. Our spiral drawing algorithm can be converted into the program

```
with        Tortoise
procedure Spiral2 is
      use Tortoise; -- results on plotter
      number_of_arms,angle_change,increment,length:INTEGER;
begin
            Get(number_of_arms);Get(angle_change);Get(increment);
            length := 0;Down;
            Loop                    -- note no case
                  exit when number_of_arms = 0;
            -- DrawArm need not appear as a nested procedure
                  length := length + increment;
                  Move(length);Turn(angle_change);
                  number_of_arms := number_of_arms - 1;
            end      Loop;
end       Spiral2;
```

ADA program: Spiral 2

Equality and inequality are not the only ways of comparing numbers; we can also use: '<' smaller than, '<=' not greater than, '>' greater than, '>=' not smaller than. The decision table

-100 < horizontal < 100	TRUE	FALSE
Draw Arm	1	
Repeat	2	

suggest a different program for drawing spirals

```
with        Tortoise
procedure Spiral3  is
        use Tortoise;    -- results on plotter
        -- provides variables horizontal and vertical
        angle_change,increment,length:INTEGER;
begin
        Get(angle_change);Get(increment);
        length := 0;Down;
        loop
                exit when horizontal > 100; -- off right side
                exit when horizontal <-100; -- off left  side
            -- why   are  vertical exits unnecessary
                length := length + increment;
                Move(length);Turn(angle_change);
        end   loop;
end     Spiral3;
```

ADA program: Spiral 3

Changing 'length := length + increment' to 'length := increment' in this spiral drawing program would be a mistake, but the same change in our other spiral drawing programs would allow them to produce polygons like

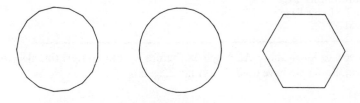

Polygons

At the age of 19 the mathematician Gauss proved that the first of these polygons can be constructed by ruler and compass, and he was so proud of his proof that his tombstone is a 17 sided polygon.

As we shall see later many kinds of values can be ordered, and we can use

$<$ before $<=$ not after
$>$ after $>=$ before

when we are designing algorithms. Consider the problem of looking up a word in a dictionary. A stupid way of solving this problem is to turn to the first page, then obey the algorithm with the decision table

Target Word $<=$ Last Word on Page	TRUE	FALSE
Find Target Word on Page	1	
Turn One Page		1
Repeat		2

The corresponding ADA named unit is

```
procedure StupidLookup is
begin
        TurnToFirstPage;
        loop
            exit when target_word
                    <= last_word_on_page;
            TurnOnePage;
        end loop;
end     StupidLookup;
```

Stupid Lookup named unit

The sensible way of looking up a word in a dictionary is given by the decision table

| Target Word $<$ First Word on Page | TRUE | FALSE | |
Target Word $>$ Last Word on Page	–	TRUE	FALSE
Find Target Word on Page			1
Turn to Earlier Page	1		
Turn To Later Page		1	
Repeat	2	2	

We need to know more ADA syntax, before we can convert this algorithm into a program. So far we have used the syntax diagrams

STATEMENT IS ALSO

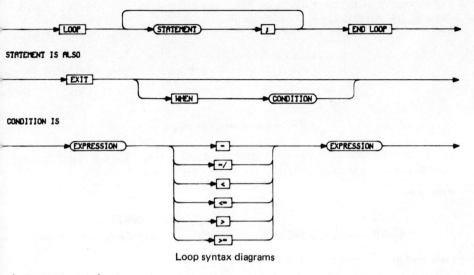

STATEMENT IS ALSO

CONDITION IS

Loop syntax diagrams

but now we need

STATEMENT IS ALSO

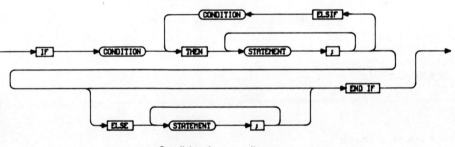

Conditional syntax diagrams

This syntax diagram is used in the named unit for the sensible way of looking up a word in a dictionary

```
procedure SensibleLookup is
begin
        TakeAnyPage;
        loop
                if    TargetWord() < FirstWordOnPage()
                then  TurnToEarlierPage;
                elsif TargetWord() > LastWordOnPage()
                then  TurnToLaterPage;
                else  exit;
                end if;
        end   loop;
end     SensibleLookUp;
```

Sensible Lookup named unit

60

Decision tables are a convenient way of specifying algorithms, but they are not as expressive as structure diagrams. When converting a decision table into a structure diagram, it is convenient to allow

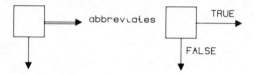

True arrows in structure diagrams

This convention allows us to convert the decision tables for our spiral drawing and dictionary lookup algorithms into the structure digrams

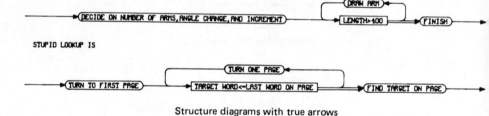

Structure diagrams with true arrows

The flexibility of the ADA *if* statements is illustrated by the structure diagrams

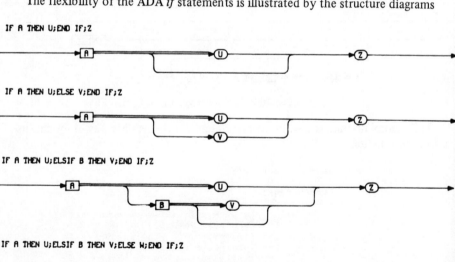

Structure diagrams for *if* statements

You can skip the rest of this section if you do not want to see the design of large repetitive algorithms for drawings arcs and lines. Suppose we want to solve the problem of drawing a quarter circle of radius r on a plotter which can only make

Approximation of a quarter circle

vertical and horizontal steps. We want an algorithm to intermingle r horizontal steps with r vertical steps, so that the plotter stays as close as possible to a true circle. Such an algorithm is given by the decision table.

Status	Above true circle	Below true circle	On true circle
No More	–	–	TRUE FALSE
Horizontal step	1		1
Vertical step		1	
Repeat	2	2	2
Finish			1

or the equivalent structure diagram

QUARTER CIRCLE IS

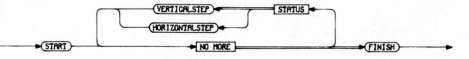

Quarter Circle structure diagram

When the plotter has made x horizontal steps and y vertical steps, it is

– above the true circle when $x^2 + (r - y)^2 < r^2$
– below the true circle when $x^2 + (r - y)^2 > r^2$
– on the true circle when $x^2 + (r - y)^2 = r^2$

This analysis shows that the problem of drawing a quarter circle can be solved by obeying the ADA program

```
with        Tortoise
procedure QuarterCircle is
      use Tortoise; --  results on plotter
      r,x,y:INTEGER;
begin
         Get(r);x:=0;y:=0;Down;
         loop
              if    x*x+(r-y)*(r-y) < r*r
              then  East;x:=x+1;
              elsif x*x+(r-y)*(r-y) > r*r
              then  North;y:=y+1;
              elsif r=y
              then  exit
              else  East;x:=x+1;
              end if;
          end  loop;
          Up;
end       QuarterCircle;
```

ADA program: Quarter Circle

Our algorithm for drawing a quarter circle can also be used for drawing a sloping line. The only change in the analysis of the problem is:

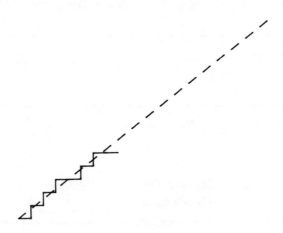

Approximation of a sloping line

if the line requires h horizontal steps and v vertical steps and the plotter has made x horizontal steps and y vertical steps, then the plotter is

- above the line when $x * v < y * h$
- below the line when $x * v > y * h$
- on the line when $\quad x * v = y * h$

Our problem of drawing a sloping line is solved by the program

```
with        Tortoise
procedure SlopingLine is
          use Tortoise;   -- results on plotter
          h,v,x,y:INTEGER;
begin
          Get(h);Get(v);x:=0;y:=0;Down;
          loop
                  if      x*v < y*h
                  then    East ;x:=x+1;
                  elsif   x*v > y*h
                  then    North;y:=y+1;
                  elsif   h+v = x+y
                  then    exit;
                  else    East ;x:=x+1;
                  end if;
          end     loop;
          Up;
end       SlopingLIne;
```

ADA program: Sloping Line

You may well be wondering why we have given this program when we could have used the Tortoise primitive Move. The analysis

West for horizontal steps North for vertical steps	East for horizontal steps North for vertical steps
West for horizontal steps South for vertical steps	East for horizontal steps South for vertical steps

suggests an ADA named unit for Move:

```
procedure Move(h,v:INTEGER) is
        -- we have given Move an extra parameter
        -- so we can avoid trigonometry
        x,y:INTEGER;
begin
        x:=0;y:=0;Down;
        loop
                if      x*v < y*h
                then    if    h < 0
                        then  West;x:=x-1;
                        else  East;x:=x+1;
                        end if;
                elsif   x*v > y*h
                then    if    v < 0
                        then  South;y:=y-1;
                        else  North;y:=y+1;
                        end if;
                elsif   h+v = x+y
                then    exit;
                else    OnLineMove;
        -- the subalgorithm OnLineMove is
        -- an exercise for the reader!
                end if;
        end     loop;
        Up;
end       Move;
```

Move procedure

When you try to devise the subalgorithm OnLineMove, you should be very careful. Almost everybody produces an algorithm which misbehaves on the axes.

Exercise

Devise an algorithm for converting from the Roman representation of numbers to the usual represenation. A crude algorithm for converting from the usual representation to the Roman representation is

ROMAN IS

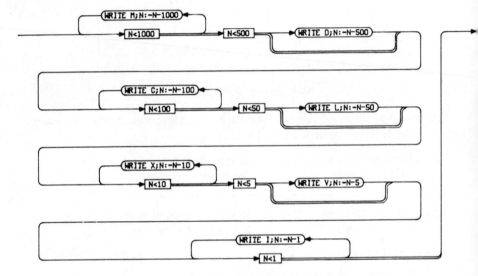

Roman structure diagram

Modify this algorithm so '90' is converted to 'XC' not 'LDDDD'.

Exercise

Some telephone companies use a system for compressing personal names. In the Soundex system the rules for compression are

- keep the first letter of the name
- for the remaining letters, drop A E I O U W H Y and convert the other letters to digits

  B C D   F G J   K L M N   P Q R S T V X Z
  1 2 3   1 2 2   2 4 5 5   1 2 6 2 3 1 2 2

- if less than 3 digits, add zeroes
- if more than 3 digits, drop superfluous digits.

EXAMPLE: COUNTESS C532, ADA A300, LOVELACE L142.

Write an algorithm for converting a name to its Soundex representation. Find two different names with the same Soundex representation. The Soundex rules are intended as a help when you want the telephone number of someone and you do not quite remember how she spells her name.

Exercise

Devise an ADA program that solves the Danish cashier's problem—to return the correct amount when a customer pays too much for something she has bought. You can assume that the Danish monetary units are 100, 50, 10, 5, 1 crowns.
  An example of a conversation with your program might be

  price: 238   customer: 30000   return: 01102

signifying that the customer has given 3 hundred crown notes for something costing 238 crowns, and the cashier should return one fifty crown note, one ten crown note and two one crown coins. How would you modify your program so it tells the cashier how many of the five monetary units remain in the till after the transaction with the customer?

Exercise

Devise an ADA program for drawing an oblong. Remember the decision table for the program Photosetter Text. Construct a decision table for a similar program that solves the problem of drawing the syntax diagrams in this book. Think about the relation between this decision table and your program for drawing an oblong.

Exercise

Devise an ADA program for printing histograms like

  A ●●●●●
  B ●●●●●●● ●● ●
  C
  D ●●

  Typical histogram

  You can assume that the data for your program consists of alternating characters and numbers like: a, 5, b, 10, c, 0, d, 2. You should check that there is room on the paper for the required number of stops.

3.3 Recursion

Sometimes the natural way to solve a problem is to use an algorithm as a subalgorithm of itself. In this section we give examples of such recursive algorithms and we explain how they can be converted into ADA programs. Our first example is artificial, but later examples will be more realistic.

Far away in the orient there is a hindu temple with piles of golden discs on three altars. At the beginning of this era two of the altars were empty, and there were 64 discs on the third altar. Each day since, the priests of the temple have taken the top disc of one of the piles and placed it on top of one of the other two piles. This era will come to an end when all 64 discs are piled on one of the altars that was empty at the beginning of the era. The problem HANOI is: which discs should the priests move each day, if they are to take us into the next era as soon as possible and they must respect the rule—never place a disc on top of a smaller disc? If the situation is

The Hanoi Problem

the priests may move the top disc of A to the top of D but not S.

The solution of the one disc problem is obvious, move the disc from S to D; the solution of the two disc problem is easy to see: move the top disc from S to A, move the last disc from S to D, then move the disc on A to D. If you try to solve the three disc problem you will probably discover the natural solution of HANOI—the case N = 64, X = S, Y = A, Z = D of the algorithm

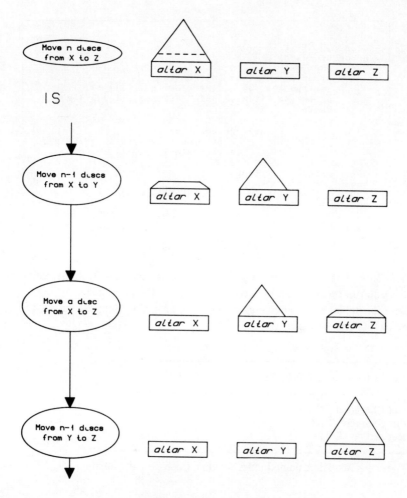

Hanoi structure diagram

The corresponding ADA named unit is

```
procedure PileMove(n:INTEGER;x,z,y:CHARACTER) is
begin
          if   n > 0
          then  PileMove(n-1,x,y,z);
                Put(x);Put("=>");Put(z);NewLine;
                PileMove(n-1,y,z,x);
          end if;
end       PileMove;
```

Hanoi named unit

Because every ADA system has an environment TEXTIO which gives a meaning
to Put and Get, we can convert this named unit into a program

```
with        Text_IO
procedure Hanoi     is
        use Text_IO;-- results on printer
        pilesize:constant INTEGER:=64;
        -- the ADA way of introducing constants
        procedure PileMove(n:INTEGER;x,z,y:CHARACTER) is
        begin
                if   n > 0
                then  PileMove(n-1,x,y,z);
                      Put(x);Put("=>");Put(z);Newline;
                      PileMove(n-1,y,z,x);
                end if;
        end     PileMove;
begin   -- Hanoi
        PileMove(pilesize,'S','A','D');
end     Hanoi;
```

ADA program: Hanoi

You should stop and think why you should replace 64 by something smaller before you run this program on a computer—an era is a long time.

Decision tables sometimes suggest recursive programs. Suppose we want to copy a line of text and we devise the decision table

Next Character is LINEFEED	TRUE	FALSE
Get Character		1
Put Character		2
Repeat		3

There are two ADA named units for this, the repetitive version

```
procedure PrintLine1 is
        c:CHARACTER;
begin
        loop
                Get(c);
                Put(c);
                exit when c = LF;
                -- LF is the line feed
                -- character signalling
                -- end of the current line
        end   loop;
end     PrintLine1;
```

Repetitive line printing

and the recursive version

```
procedure PrintLine2 is
        c:CHARACTER;
begin
        Get(c);
        Put(c);
        if    c /= LF
        then  PrintLine2;
        end if;
end     PrintLine2;
```

Recursive line printing

Because there are devices that can print lines from right to left as well as left to right, we might want to modify our decision table

Next Character is LINEFEED	TRUE	FALSE
Get Character		1
Put Character		3
Repeat		2

This decision table gives an algorithm for printing a line backwards; it is difficult to convert it to a repetitive named unit, but it is easy to convert it to a recursive named unit

```
procedure ReversePrint is
      c  :CHARACTER;
begin
        Get(c);
        if    c /= LF
        then    ReversePrint;
        end if;
        Put(c);
end     ReversePrint;
```

Reverse printing

We can use our algorithms Print and Reverse Print in an algorithm for printing lines efficiently.

```
with        Text_IO
procedure FastPrint is
      use Text_IO;   -- results on reversible printer
                     -- LF changes print direction
      c  :CHARACTER;

      procedure Forwards is
      begin
                  Get(c);Put(c);
                  if    c /= LF
                  then  Forwards;
                  end if;
      end         Forwards;

      procedure Backwards is
      begin
                  Get(c);
                  if    c /= LF
                  then  Backwards;
                  end if;
                  Put(c);
      end         Backwards;

begin   -- FastPrint
        loop
                  Forwards;
                  Backwards;
            end   loop;
        -- soon you will learn how the
        -- machine leaves such a loop
      end         FastPrint;
```

ADA program: Fast Print

Some programmers prefer recursive algorithms to repetitive algorithms and they exploit the fact that recursion can always replace repetition. Find out your preference by comparing the repetitive Move in the last section with

```
procedure Move(h,v:INTEGER) is

      procedure DrawLine(x,y:INTEGER) is
      begin
                  if     x*v < y*h
                  then   if    h < 0
                         then  West;DrawLine(x-1,y);
                         else  East;DrawLine(x+1,y);
                         end if;
                  elsif  x*v > y*h
                  then   if    v < 0
                         then  South;DrawLine(x,y-1);
                         else  North;DrawLine(x,y+1);
                         end if;
                  end if;
      end         DrawLine;

begin   -- Move
        Down;DrawLine(0,0);Up;
      end         Move;
```

Recursive Move procedure

When the natural algorithm for solving a problem uses itself more than once as a subalgorithm, you should always use recursion in the first version of your program. Suppose we want to draw pictures like

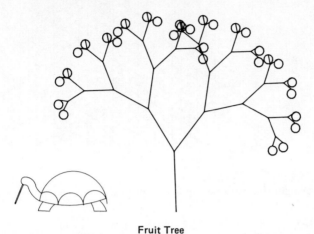

Fruit Tree

The natural algorithm for solving this problem is

TREE (N, DIRECTION) IS

Fruit Tree structure diagram

which we can convert into the program

```
with        Tortoise
procedure FruitTree is
      use Tortoise; -- results on plotter
      age:constant INTEGER := 6;

      procedure Tree(n,direction:INTEGER) is
      begin
                  if    n = 0
                  then  DrawApple;-- left to reader
                  else  Tree(n-1,direction-30);
                        Turn(300);
                        Tree(n-1,direction+40);
                  end if;
      end         Tree;

begin  -- FruitTree
            Down;Tree(age,90);Up;
end         FruitTree;
```

ADA program: Fruit Tree

We can use a variation of this program to draw

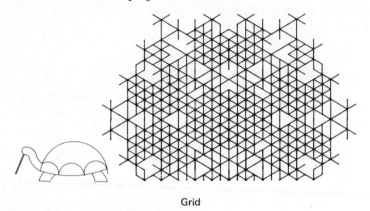

Grid

The natural algorithm for drawing such figures is

GRID(N,DIRECTION) IS

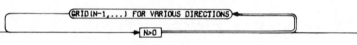

Grid structure diagram

which we can convert into the program

```
with        Tortoise
procedure Grid is
        use Tortoise; -- results on plotter
        age:constant INTEGER := 6;

        procedure Subgrid(n:INTEGER) is
        begin
                if     n > 0
                then   loop
                            Move(20);Turn(60);
                            Subgrid(n-1);
                            exit when direction=360;
                      end    loop;
                end if;
        end        Subgrid;

begin    -- Grid
        Down;Subgrid(age);Up;
end      Grid;
```

ADA program: Grid

Unfortunately this program is inefficient, because some lines are drawn several times. When you write a recursive program you should be on the lookout for inef-

73

ficiency; if your time is less valuable than computer time, you can write a more
complicated program that avoids inefficient recursion.

Exercise

Consider patterns of black and white squares like

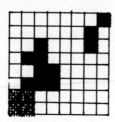

Connectivity problem

Devise an algorithm the number of squares in the connected black area containing a given square. Show that, if you give your algorithm the above pattern and the
squares (1,1), (3,3), (2,7), it will return the numbers 0, 12 and 2.

Exercise

On the display keyboard you cannot type the sequence of symbols:
Note the *nesting αβγδ here*; instead you can type a sequence of symbols like

(I) Note the _n_e_s_t_i_n_g %a%b%c%d_h_e_r_e
(II) Note the [−nesting] [%abcd] [_here]
(III) Note the [_nesting [%abcd] here]

Devise an algorithm for converting sequences in form III to sequences in form
I and II. Devise algorithms for printing sequences in forms I, II and III on the
assumption that you have subalgorithms Normal, Underline and Greek for printing
in various fonts. As font changes are rare, most photosetters save space by using
II or III instead of I. If you have used the lock on the case shift key of a typewriter,
you have used II instead of I.

Exercise

Devise an algorithm for finding the number of paths in the diagram

V
O O
L L L
A A A A
P P P
Y Y
K

that spell **VOLAPYK** (the name of a universal language invented by a Dane as a rival to Esperanto). Modify your algorithm so that it computes the probability of EGO beating ID in a game of table tennis when the probability of EGO winning any particular point is p. What changes would you make if the game had been badminton or tennis? Which scoring system do you think is best?

Exercise

Examine carefully the structure of the Hilbert curves

Space filling curve

then write an ADA program that draws them. (If you enjoyed this exercise, you can find many similar curves in B. B. Mandelbrot, *Fractals: Form, Chance and Dimension*, W. H. Freeman 1977.)

3.4 Exceptions and errors

Hitherto our way of solving problems has been: to state the problem, devise an algorithm on the assumption that we have various subalgorithms, design the assumed subalgorithms, then convert the algorithms and subalgorithms to an ADA program. In this section we focus on the questions

— how can we be sure that an ADA program solves a given problem?
— how should a problem solution be documented?

Documentation is important; an undocumented program is unintelligible and it is unmaintainable because even its designer will soon forget what it is supposed to do. Remember our solution of the problem of drawing a quarter circle. When we give a reasonable number to the program

```
with        Tortoise
procedure QuarterCircle is
        use Tortoise; --   results on plotter
        r,x,y:INTEGER;
begin
        Get(r);x:=0;y:=0;Down;
        loop
            if     x*x+(r-y)*(r-y) < r*r
            then  East;x:=x+1;
            elsif x*x+(r-y)*(r-y) > r*r
            then  North;y:=y+1;
            elsif r=y
            then  exit
            else  East;x:=x+1;
            end if;
        end  loop;
        Up;
end     QuarterCircle;
```

ADA program: Quarter Circle

we do in fact get a reasonable quarter circle, but this is not enough. If we want to rely on this program as a solution of the quarter circle problem, if we want to give it to our friends or use it in solutions of other problems we need to know more— not only what values of r are reasonable, but also why the program is as it is. Even although our description of the solution of the quarter circle problem in section 3.2 had an analysis of the problem, and the program algorithm in the form of a structure diagram, this description is not a satisfactory documentation of our problem solution because we did not say (1) negative values of r are unreasonable, (2) the plotter will go over the edge of the paper if the value of r is too large, (3) the program was designed this way, because we wanted it to be easy to understand. We would have had a satisfactory documentation if we have given this information and a description of the names used in the program:

r   ——   radius of the desired quarter circle
x   ——   the number of horizontal steps the plotter has made
y   ——   the number of vertical steps the plotter has made
East ——   a Tortoise primitive for making a horizontal step
North——   a Tortoise primitive for making a vertical step.

This documentation is satisfactory because it enables us to understand and modify the problem solution.

Suppose we decide to make our solution of the quarter circle problem more robust by checking that the value given to r is reasonable, and we decide to make our solution more efficient by not squaring (r-y). We can achieve these aims by changing the structure diagram to

QUARTER CIRCLE IS

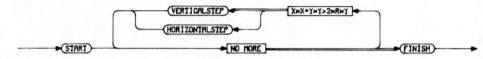

Revised Quarter Circle structure diagram

and the ADA program to

```
with        Tortoise
procedure QuarterCircle is
       use Tortoise; -- results on plotter
       r,x,y:INTEGER;
       Unreasonable:EXCEPTION;
begin
       x := 0;y := 0;Get(r);
       if r <    0 then raise Unreasonable end if;
       if r > 100 then raise Unreasonable end if;
       loop
              exit when r = y;
              if     x*x+y*y > 2*r*y
              then   North;y := y+1;
              else   East ;x := x+1;
              end if;
       end    loop;
exception
       when Unreasonable => Put("Unreasonable Radius");
end       QuarterCircle;
```

Revised ADA program: Quarter Circle

In this program we have used an ADA exception to interrupt the normal flow of control; if the user gives the program bad data it will raise an exception instead of trying to draw a quarter circle.

You will learn more about exceptions if we use them in a solution of the problem of converting unjustified text like

```
Der kom en soldat marcherende henad
landevejen: Een, To. Een, To. Han
havde sit tornyster på ryggen og en
sabel ved siden, for han havde været i
krigen, og nu skulle han hjem. Så mødte
han en gammel heks på landevejen; hun
var så ækel, hendes underlæbe hang hende
lige ned på brystet. Hun sagde: God
aften soldat, hvor du har en pæn sabel
og et stort tornyster,du er en rigtig
soldat. Nu skal du få så mange penge,
du vil eje.
```

Unjustified Text

to justified text like

```
Der   kom   en   soldat   marcherende   henad
Landevejen:    Een,    To.   Een,    To.   Han
havde   sit   tornyster   på   ryggen   og en
sabel   ved   siden, for han havde været i
krigen,   og nu skulle han hjem. Så mødte
han   en   gammel   heks  på   Landevejen; hun
var så ækel, hendes underlæbe hang hende
Lige   ned   på   brystet.   Hun   sagde: God
aften   soldat, hvor du har en pæn sabel
og   et   stort   tornyster,du er en rigtig
soldat.   Nu   skal   du få så mange penge,
du                       vil                      eje.
```

Justified Text

As the documentation of any problem solution should contain

— a specification of the problem
— a precise description of the algorithm and subalgorithms underlying the solution
— the design decisions behind the algorithm
— the program itself and a description of its use of names
— a user guide, revealing the requirements on the program data and the significance of program results

we will provide this information for our justification problem.

The specification 'devise an ADA program for aligning the left and right margins of a text' is adequate for readers who know about aligning and margins, but the example we gave in the last paragraph might well be included in the specification also. The next part of the documentation of our solution is a precise description of the algorithms and subalgorithms:

JUSTIFY IS

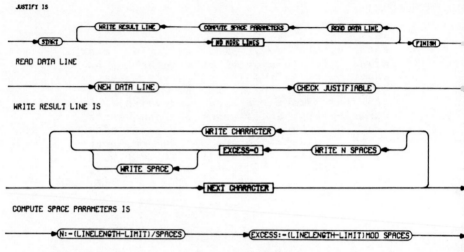

READ DATA LINE

WRITE RESULT LINE IS

COMPUTE SPACE PARAMETERS IS

Justify structure diagrams

As the third part of the documentation we should record the design decisions (1) to avoid linguistic problems like the division of long words (2) to compute two numbers, n and excess, such that: linelength= nonspaces + (n+1) × spaces + excess (3) to check for exceptional situations:

```
Too Long Line  -- more than 40 characters in the line
No Spaces      -- none of the  characters in the line
               -- is a  space so we cannot justify.
```

Justify exceptions

The fourth part of the documentation might be the ADA program

```
with      LineManipulator
   -- a description of this environment
   -- can be found in the next chapter
procedure Justify is
         use LineManipulator; -- results on printer
         l  :LINE;c:CHARACTER;
         line_length:constant INTEGER := 40;
         n,excess:INTEGER;
         TooLongLine,NoSpaces,NoMoreData:EXCEPTION;

         procedure ReadDataLine is
         begin
                  l := DataLine();
                  if    limit = 0
                  then  raise NoMoreData;
                  elsif limit > line_length
                  then  raise TooLongLine;
                  elsif spaces= 0
                  then  raise NoSpaces;
                  end   if;
         end      ReadDataLine;

         procedure ComputeSpaceInsertionParameters is
         begin
                  n     :=(linelength-limit) / spaces;
                  excess:=(linelength-limit) mod spaces;
         end      ComputeSpaceInsertionParameters;

         procedure WriteResultLine is
              sofar:INTEGER;
         begin
                  loop
                      c := NextCharacter();
                      case c is
                          when ' '   => sofar:=0;
                                        loop
                                            sofar  := sofar+1;
                                            Put(' ');
                                            exit when sofar=n;
                                        end loop;
                                        if    excess > 0
                                        then  excess:= excess-1;
                                              Put(' ');
                                        end if;
                          when LF    => NewLine;exit;
                          when others => Put(c);
                      end case;
                  end loop;
         end      WriteResultLine;

begin   -- Justify
         loop
                  ReadDataLine;
                  ComputeSpaceInsertionParameters;
                  WriteResultLine;
         end   loop;
         -- Left and right margins aligned
exception
         when TooLongLine => Put("Too Long Line");
         when NoSpaces    => Put("Can't Justify without Spaces");
         when others      => null;-- NoMoreData is usual exit
end      Justify;
```

ADA program: Justify

Note the ADA declarations of exception names, the ADA statements for raising the exceptions, and the ADA statements for handling exceptions when they arise

STATEMENT IS ALSO

DECLARATION IS ALSO

BODY IS ALSO

Exception syntax diagrams

Normally our program Justify repeatedly obeys the procedures Read Data Line, Compute Insertion Parameters, Write Result Line until there are no more lines to justify. However data that raises an exception will make the program send an error message and die. When an exception is raised, the normal flow of control is abandoned and the program looks for a handler for this exception. If the exception No Spaces arises in Read Data Line and Read Data Line was entered from Justify, then the program will (1) look for 'when No Spaces =>' or 'when others =>' in ReadDataLine (2) if unsuccessful, look for them in Justify (3) if unsuccesful, die.

Let us complete the documentation for our solution of the justification problem. We must provide a description of the names used in the program:

l	is a variable for keeping the line being justified
spaces	is a variable for counting the blank characters
nonspaces	is a variable for counting the other characters
n	tells how many blanks replace a single blank
excess	tells how many extra blanks must be inserted
sofar	is an auxiliary variable
DataLine	fetches a data line from the keyboard
NextCharacter	fetches a character from a data line
NoMoreLines	is true when all the data has been justified.

We must also provide a user guide:

The ADA program Justify aligns the left and right margins of a text. If some text line is longer than 80 characters or there is no blank in some text line, the program produces an error message and dies. Remember to translate the environment Line Manipulator before translating Justify.

Now that we have completed the documentation we must make sure that it documents a program that does solve the problem it is supposed to solve.

In the last chapter we described the pessimist's way of running a program on the computer. Pessimists are prepared for exceptional situations, for errors in their programs and data. Program errors can be divided into:

— Runtime errors which are trapped by the computer algorithm DO because they raise an exception.
— Translation time errors which are trapped by the computer algorithm Translate
— Conceptual errors that make your program solve a different problem from the one it is supposed to solve.

We shall describe later how one can repair program errors and how one can detect conceptual errors. Because the computer can catch translation and run time errors, they are not so disastrous. Even so, most of the people at your local computer centre spend their time repairing program errors. It has been said that run-time errors can occur in all large programs. You should know how to 'debug' your programs. One useful principle is to run your program with well chosen test data. If your test data is such that every line in your program is checked *and* it checks all exceptional situations that can arise, then it is probably well chosen and it should be part of the program documentation. We suggested putting the unjustified and justified texts that were the test data for Justify in the specification part of the documentation but we would not quarrel with one who chose to put them in the user guide part.

We should say a little about the exceptions raised by run time errors. Consider the program

```
with        Text_IO
procedure GoldenSection is
        use Text_IO;
        a,b:INTEGER;
begin
        a:=1;b:=1;
        loop
                a:=a+b;
                b:=a-b;  -- b gets a's old value
        end  loop;
exception when Numeric_Error =>Put(1000000*a/b);
end        GoldenSection;
```

ADA program: Golden Section

When the value of an integer expression is too large for the computer, it raises the predefined exception Numeric-Error. This exception and the other predefined exceptions we shall meet later are handled in the same way as user defined exceptions.

The number generated by this program, divided by a million, is the so-called golden section ϕ; it fascinates many people because it turns up in the most surprising places—e.g. many consider a room to be well proportioned if its length divided by its breadth is near to ϕ.

The syntax diagram

STATEMENT IS ALSO

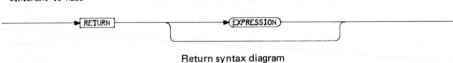

Return syntax diagram

gives a very useful way of breaking the normal flow of control; it sends control to the end of the current procedure in the same way that an exit statement sends control to the end of the current loop. If a return statement contains an expression, then the value of the expression is the value returned by the function. Because the return statements are the only way a function can return a value, you will find yourself writing many of them in your ADA programs. Let us close this section with an example

```
with        Text_IO
procedure GoldenSection is
        use Text_IO; -- results on printer
        function Fibonacci(n:INTEGER) return INTEGER is
        begin
                if    n < 2
                then  return 1;
                else  return Fibonacci(n-1)+Fibonacci(n-2);
                -- why is this inefficient?
                -- improve it!
                end   if;
        end     Fibonacci;
begin   -- GoldenSection
        Put(1000000*Fibonacci(20)/Fibonacci(19));
end     GoldenSection;
```

Revised ADA program: Golden Section

Exercise

How does our program Justify react in other exceptional situations that can occur when texts are being justified: initial spaces, final spaces, successive spaces? Design a version of Justify that behaves more intelligently in these situations. If your program sends an error message and dies when an exceptional situation arises, design a version that does not die.

Exercise

Modify some of the programs in this chapter so they use *return* instead of *exit*. Modify some of the programs you have wtitten earlier so they use exceptions, then write a user guide for them.

Exercise

If you write ADA programs for drawing

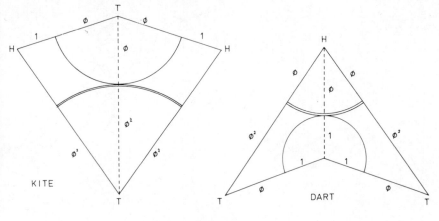

Kite and Dart

and run them many times, you can make pretty patterns (read more about them in the January 1977 issue of *Scientific American*).

Kite-Dart pattern

3.5 Streaming, pipelining and tasks

In this section we describe how the computer can do many things at the same time, and how an ADA program can take advantage of this. Let us begin with a situation where the computer does not do many things at the same time. When we ask the computer to obey

Normal program execution

it takes characters and numbers from 'data' when it meets Get in the program; it puts characters and numbers in 'results' when it meets Put in the program; information flows in a stream from 'data' to 'results'.

A situation that often arises in practice is: program A generates results that are used by program C and never used again. In most computer systems one has to use a temporary file, one has to ask the computer to obey

$\rightarrow$ (DO(A, INPUT, LAKE)) $\rightarrow$ (DO(C, LAKE, OUTPUT)) $\rightarrow$

Streaming

If the computer can work on A and C at the same time, it need not save the intermediate results on a temporary file, C can use an intermediate result as soon as it is produced by A. In some computer systems one can indicate this possibility of streaming by asking the computer to obey Do(C < = A, input, output) instead of Do(A, input, lake) followed by Do(C, lake, output). In such computer systems one can build long pipelines

Pipelining

When we ask the computer to obey

$$Do(C < = B_m < = \ldots < = B_2 < = B_1 <= A, \text{Data, Results})$$

we get the same result as when we ask it to obey Do(A, Data Temp 1), then Do(B_1, Temp 1, Temp 2) . . . The advantage of pipelining is that the computer can run A, $B_1, B_2, \ldots, B_n$, C in parallel instead of wasting space on temporary files.

Concurrency, otherwise known as parallelism is becoming more and more important as multiprocessors and computer networks become more widespread. For many years most large computers have supported simultaneous users, and the architecture of modern computers gives more and more support to concurrency. Programming languages have followed this development, and, like ADA, many now support streaming, pipelining and other forms of concurrency.

Imagine the computer as an actor, the algorithm as the actor's script, and the computer obeying an algorithm as the actor performing the actions prescribed by the script. Real actors can play many roles; at the will of the director they will play the role of Hamlet, Lear or someone else. Corresponding to an actor prepared to play several roles, we have the ADA concept of a *task* prepared to accept several *entries*. In the course of a season an actor may play zero, one or more roles; in obeying a program a task may accept zero, one or more of its entries. If we want to build the pipeline $C < = A$, we can program C as a task with just one entry: Blow. When A generates something which C can work on, it can obey 'Blow (something)' and we may have both A and C active. We would have had a result-driven pipeline instead of a data-driven pipeline, if we had programmed A as a task with just one entry: Suck. When C wants something from A it can obey 'Suck (something)', and again we may have both A and C active at the same time. Perhaps you noticed that ADA statements for calling an entry are indistinguishable from ADA statements for calling a procedure. Because of this we need not change the program C in our result-driven pipeline, if we decide to abandon concurrency and make Suck the name of a procedure.

Actors are like ADA tasks

As another example of concurrency let us describe how the computer can separate the data and results of several programs that are running at the same time. How can the computer avoid mixed up drawings when you ask it to obey

Do(yourcode, your data, plotter)

at the same time that it is obeying

Do(hercode, her data, plotter)

for someone else? The answer might be:

— while yourcode and hercode are being obeyed, results are put on temporary files and nothing is plotted;
— when yourcode and hercode are finished, they call an entry in the plotter task;
— when the plotter task accepts an entry call, it converts a temporary file to a drawing.

The reader can assume that there is an **ADA** task for each output device on her computer, and output appears at the device when the device task accepts the call of an entry.

Because parallelism is not easy to understand, we will introduce the concept of a network diagram and redo our examples in great detail. Network diagrams look like structure diagrams but their interpretation is quite different. A network diagram consists of oblongs joined by arrows

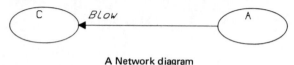

A Network diagram

Each oblong in the diagram represents a task; at any time each oblong is in one of the states:

GREEN — active, doing useful work
(our actor is playing a role)
YELLOW — waiting for some other task to call one of its entries
(our actor is unemployed)
RED — waiting because it has called an entry in some other task
(we have a role for some actor)
WHITE — passive
(our actor is in some distant land).

Initially all oblongs are WHITE but they can change colours according to the rules:

— a GREEN oblong can change colour at any time;
— a WHITE oblong can change into a GREEN oblong;
— if we have an arrow from a RED oblong to a YELLOW oblong, then we may have a *rendezvous* and both oblongs become GREEN;
— if several rendezvous are possible with a YELLOW oblong, then only one of them can actually happen.

Notice that these rules never say which colour changes must occur, only which colour changes may occur.

In our examples mysterious black dots will appear and disappear as the oblongs in the network change colour. Later we shall explain the significance of the black dots and the subtle difference between WHITE and GREEN oblongs. Consider the initial network diagram for our data-driven pipeline $C < = A$

initial network diagram

Suppose A and C become active at the same time

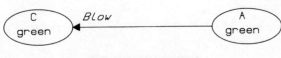

when A C become active

then the task C waits for A to call its entry Blow

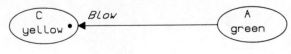

when C is waiting for A

then A calls the entry Blow

when A is ready

then the task C accepts the entry call

after the rendezvous

The last colour change was given by the rendezvous rule. In this example the owner of an entry waited for the entry to be called; in the next example the entry caller waits because the entry owner is not ready to accept the call. The initial network diagram for our result-driven pipeline C < = A is:

another initial network diagram

One of the possible sequence of colour changes is A and C become active at the same time:

When A C become active

then C calls the entry Suck

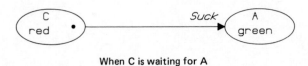

When C is waiting for A

then A is ready to accept Suck

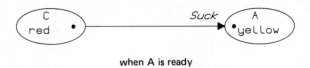

when A is ready

then a rendezvous occurs

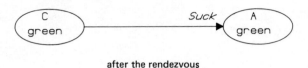

after the rendezvous

and both A and C can be active at the same time.

Our last example is your asking the computer to obey Do(yourcode, yourdata, plotter) when it is obeying Do(hercode, herdata, plotter) for someone else. This gives the network diagram

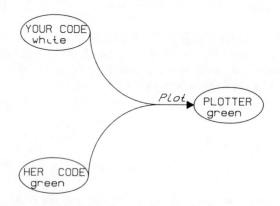

Plotter network diagram

Three events can change the colours of these oblongs

— the computer may start work on yourcode
— the computer may finish work on hercode
— the plotter may ask for more work.

If all three events happen at the same time, we get

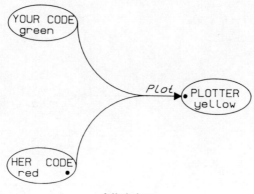

A little later

Two events can change these colours

— the computer may finish work on yourcode
— the plotter may make a rendezvous with hercode.

If both events happen at the same time, we get

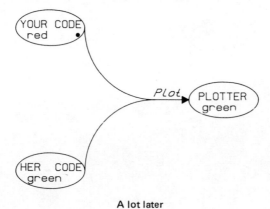

A lot later

and the results of hercode will be plotted. If only the first event had happened, we
would have had

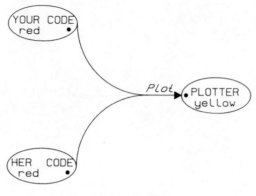

Another possibility

Two events can change these colours

— the plotter may make a rendezvous with yourcode
— the plotter may make a rendezvous with hercode

but the colour change rules ensure that these two events *cannot* occur at the same time. If the first event were to occur, we would get

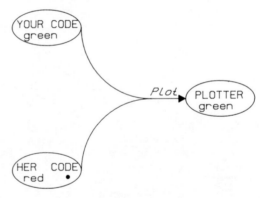

After the rendezvous

and the computer would plot the results of yourcode *before* the results of hercode, even although you asked it to obey Do(yourcode, yourdata, plotter) *after* it was asked to obey Do(hercode, herdata, herdata, plotter). There is an ADA rule which ensures that this cannot happen for our sequence of colour changes—if a call of an entry is made before another call of the same entry, then the first call is accepted before the second. Note that your plot will always appear before her plot if the computer finishes yourcode before it finishes hercode.

We will tell you how to write an ADA task in a later chapter. Until then you will only need to know that (1) entry calls look exactly like procedure calls (2) you can assume tasks are never WHITE (3) a task may be waiting for many different entries at the same time. The mysterious black dots in our network diagram indicate which entries tasks are waiting for. A YELLOW oblong in a diagram has one or more dots on the heads of entering arrows and no dots on the tails of leaving arrows;

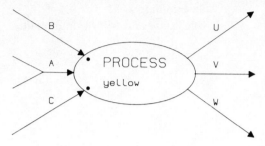

A task can wait for several entries

a RED oblong in a diagram has no dots on the heads of entering arrows and one dot on the tail of a leaving arrow

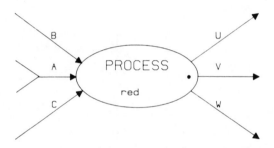

A task can only call one entry

Note that there is one arrow head for each entry a task owns, and one arrow tail for each entry a task calls.

Exercise

We can have network diagrams where all oblongs are RED or YELLOW and no rendezvous is possible. This situation is known as deadlock; it fascinates some computer scientists; it sometimes causes chaos in the computer solution of practical problems. Use network diagrams to describe two people wanting to go through a door at the same time. Did you describe a collision or an 'after you—no after you' deadlock?

Exercise

Because thinking and typing take much more time than editing, we want the computer to obey Edit(yourdata, your result) and Edit(herdata, her result) at the same time. Design Edit as a task with three entries

Start(data)   — which copies data into an edit file;
OneLine(corrections)— which reads one line of corrections and modifies the edit
                           file accordingly;
Finish(result)   — which copies the edit file into result.

Chapter 4

Environments

The best way of solving a large problem is divide-and-conquer; to assume you have an environment in which many subproblems have been solved and combine these solutions into a solution of the large problem. You have seen programs that use the environment Tortoise to make intricate drawings, but you have not learnt how environments are made. When you use an environment, you need to know which primitives the environment provides, but you do not care how these primitives are realized. In the first section of this chapter we describe the ADA way of telling the user of an environment what she wants to know about its primitives; in the remaining sections we present several useful environments to illustrate the slogan 'careful design of environments is the key to solving large problems in ADA'.

4.1 Interfaces and parameters

The ADA way of telling the user of an environment about its primitives is to give an *environment interface* like

```
package Geometer is
        procedure  Point;
        procedure  Square;
        procedure  Triangle;
end     Geometer;
```

Geometer package interface

The user sees that the environment Geometer provides three parameterless procedures Point, Square and Triangle; she assumes that the designer of Geometer will provide an *environment body* which defines these procedures, and she does not care about the details in these procedure definitions.

The interface for the environment Tortoise is

95

96

```
package Tortoise is
        -- results on plotter
        penstatus,direction:INTEGER;
        horizontal,vertical:INTEGER:=0;
        procedure Up;
        procedure Down;
        procedure North;
        procedure South;
        procedure East;
        procedure West;
        procedure Move(n:INTEGER);
        procedure Turn(d:INTEGER);
        procedure TurnTo(d:INTEGER);
end     Tortoise;
```

Tortoise package interface

We see that the environment Tortoise provides 9 procedures and 4 variables. You have already met the variables pen, status and direction in the form

Direction and Pen Status variables

You also met the variable horizontal in the program Spiral 3 for drawing a spiral without running the risk of the pen moving over the edge of the paper.

Environment interfaces can not only provide users with convenient procedures and variables, but they can also provide useful concepts. The interface

```
package LineManipulator is
        type LINE is new STRING(1..80);-- explained later
        limit,spaces:INTEGER;
        function DataLine             return LINE;
        -- always a keyboard line
        function LineToCharacter return CHARACTER;
        function NextCharacter     return CHARACTER;
        procedure Insert(l:LINE);
end     LineManipulator;
```

Line Manipulator package interface

gives the concept LINE to the user. The user can introduce variables whose values are lines; the program Justify in section 3.4 kept a line in such a variable while it was doing its space computations.

You may well have been worried by 'variables whose values are lines' in the last sentence. Your worry may disappear when we philosophise a little on the analogous 'variables whose values are numbers'. Europeans represent positive whole numbers by symbol sequence like

12   5   1980

while the romans would have represented the same positive whole numbers by quite different symbol sequences

XII   V   MCMLXXX

Everybody uses representations of numbers in their daily life, but nobody knows what numbers 'are'. Analogously we can use representations of lines in algorithms without knowing what lines are. Consider the Tortoise drawing

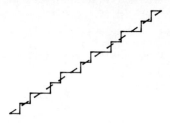

Approximation to sloping line

This drawing is a representation of a sloping line that can also be represented as the symbol sequence

eneneeneneeneneeneneneenene

If you run some of our Tortoise programs on two different computers, one of them may give you a drawing, while the other may give you a symbol sequence because it does not have a plotter. In this chapter we will make no assumptions about the representation of values. When we say that a drawing is the result of running a program, we mean: the drawing is one representation of the result, but there may well be other representations.

Computers spend much of their time converting from one representation of values to another. The program Sloping Line in section 2.4 converted a pair of numbers into a sloping Tortoise line. You may have liked the Tortoise package better if it had used Sloping Line instead of Turn, TurnTo, Move . . .

```
package SimpleTortoise is
       -- results on plotter
       procedure Down;
       procedure Up;
       procedure MoveTo(x,y:INTEGER);
       -- remember SlopingLine in section 2.4
   end    SimpleTortoise;
```

Simple Tortoise package interface

Where is the best place to convert a pair of numbers into a sloping line? You have seen two possibilities—in a special program, in an environment; two other possibilities are—in tne task for plotter output, in a microprocessor in the plotter itself. The moral of this paragraph is that conversion programs are tedious, you should not write them until they cannot be avoided. Even if you do not find conversion programs boring, you should follow this advice because most conversion programs are expansion programs. Bloated information wastes time when it has to be moved, it wastes space in the computer, it costs money.

Let us look at an example of the three ways an environment interface can introduce a new type instead of worrying about possible representations of its values:

```
(PUBLIC)   package DateManipulator is
                   type DATE is new INTEGER;
                   function Create(day,month,year:INTEGER) return DATE;
                   function Interval(after,before:DATE) return INTEGER;
           end     DateManipulator;

(PRIVATE) package DateManipulator is
                   type DATE is private;
                   function Create(day,month,year:INTEGER) return DATE;
                   function Interval(after,before:DATE) return INTEGER;
          private type DATE is new INTEGER;
          end     DateManipulator;

(LIMITED) package DateManipulator is
                   type DATE is limited private;
                   function Create(day,month,year:INTEGER) return DATE;
                   function Interval(after,before:DATE) return INTEGER;
          private type DATE is new INTEGER;
          end     DateManipulator;
```

Public, private and limited packages

When an environment interface introduces a type T, the user can always have values and parameters of type T. If T was introduced in the limited way, the user cannot assign a value to a variable of type T; if T was introduced in the private way, the only values a user can assign to a variable of type T are those generated by the environment primitives. If the user of the second version of Data Manipulator has DATE variables: birthday, age, today; she can write assignments like

birthday := Create(4,12,1937);
age   := Interval(today, birthday);

but she cannot write: birthday := 4 × 12 × 1937. The user of the first version of Date Manipulator can write all three assignments, because there are no restrictions on how values of values of a new type are generated when the new type is introduced in the public way. From now on we shall never introduce types in the public way because we want all our programs to work for all our readers, also those blessed with environment bodies that are not the same as those we shall give to Tortoise and our other environments.

Our design of Date Manipulator reflects the decision that we prefer 'age := Interval(today, birthday)' to 'age := today-birthday' because subtracting dates is a more complicated affair than subtracting numbers. If we had decided that this difference was less important than the convenience of '-' and the usual infix notation of mathematics, we would have given the module interface

```
package DateManipulator is
        type DATE is private;
        function Create(day,month,year:INTEGER) return DATE;
        function "-"   (after,before:DATE)   return INTEGER;
private type DATE is new INTEGER;
end     DateManipulator;
```

Overloaded function

and the usual meaning of '-' would have been overloaded with a new meaning: subtraction of dates.

The environment interfaces you have seen so far have only used one of ADA's three parameter mechanisms, that given by the first row in the table

	value read when procedure starts	value rewritten when procedure stops
in parameter	YES	NO
inout parameter	YES	YES
out parameter	NO	YES

Consider the interface

```
package DateManipulator is
        type DATE is private;
        procedure Create(day,month,year:in INTEGER;d:out DATE);
        procedure Interval(after,before:DATE;n:out INTEGER);
        -- in assumed if out or in out not written
        procedure OneDayMore(d: in out DATE);
private type DATE is new INTEGER;
end      DateManipulator;
```

Date Manipulator package interface

Instead of 'birthday := Create(4, 12, 1937)' the user must now write 'Create(4, 12, 1937, birthday)' and the procedure Create

— takes 4, 12, 1937 as the initial values of day, month, year
— copies the final value of d into birthday

Instead of 'age := Interval(today, birthday)' the user must now write 'Interval(today, birthday, age)' and the procedure Interval

— takes the values of today and birthday as the initial values of after and before
— copies the final value of d into age.

If the user writes 'One Day More(today)', the procedure One Day More

— takes the value of today as the initial value of d
— copies the final value of d into today.

Try to develop a balanced programming style; sometimes *in, out, inout* parameters are convenient, but judicious use of *return* often gives a more readable program.

Exercise

Reformulate the interface for Line Manipulator so that it uses *out* instead of *return*. Reformulate the program Justify in section 3.4 so that it uses the *out* version of Line Manipulator.

4.2 The environment Drawing

For our first realistic example of environment design we consider the problem of drawing the line-and-arc figures in this book

Family drawing

Such a figure is a value of the concept FIGURE which we shall introduce in our environment Drawing. More precisely: if we obey the environment primitive PutFigure(f) in a program which sends its results to the plotter, then the value of f will be drawn. As we want to keep figures in the computer from one program run to another, our environment will also have a primitive Get Figure(f) which fetches a figure from its data and assigns the figure to the variable f.

The next step in the design of the environment Drawing is to decide on the primitives for drawing very simple figures. If we decide on sloping lines, we can have a primitive Line specified by the test data set

Line (10, 0)
Line (7, 45)
Line (14, 135)

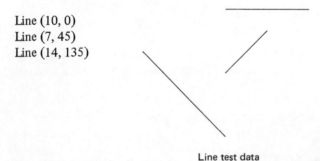

Line test data

If we decide on arcs we can specify a primitive Arc by the test data set

Arc (10, 0, 180)

Arc (7, 45, 0)

Arc (14, 135, 90)

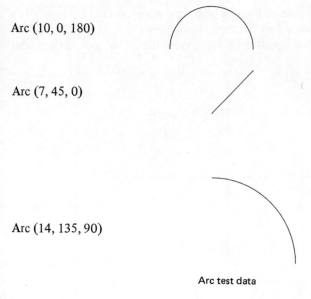

Arc test data

As we want to make more complicated figures than lines and arcs, our environment Drawing will have a primitive Catenate with the specification

Catenate test data

102

In this specification we assume that the lines separating rows and columns give a frame for the figures positioned within them. Remember this when you look at the specifications for the other primtives—Shift, Scale, Rotate—in our environment Drawing:

Shift test data

Scale test data

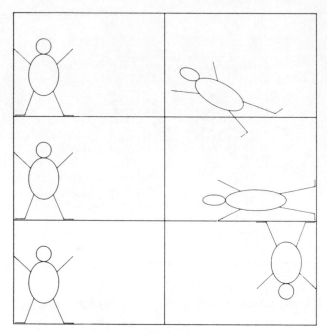

Rotate test data

The user's view of the environment Drawing is given by the primitive specifications and the module interface:

```
package Drawing is
       -- results on plotter
       type FIGURE is private;
       function Line      (length,direction:INTEGER)   return FIGURE;
       function Arc(length,direction,angle:INTEGER)    return FIGURE;
       function Catenate      (left,right  :FIGURE )   return FIGURE;
       function Shift (f:FIGURE;length,direction:INTEGER)return FIGURE;
       function Scale     (f:FIGURE;procent:INTEGER)   return FIGURE;
       function Rotate    (f:FIGURE;angle:INTEGER)     return FIGURE;
       function GetFigure                              return FIGURE;
       procedure PutFigure(f:FIGURE);
private  -- FIGURE specified later
end      Drawing;
```

Drawing package interface

If we want the first drawing in this section, then we can obey the program

104

```
with       Drawing
procedure Marriage is
      use Drawing; -- results on plotter
          man,woman,couple,torso,head,arms,legs:FIGURE;
begin  --  discover where we have cheated,
       --  the created figures are not those promised
          torso:= Arc(40,90,360);
          head:= Shift(Arc(15,90,360),40,90);
          arms:= Shift(Line(20,45),28,45) &
                 Shift(Line(20,135),28,135);
          legs:= Shift(Line(30,290),10,10) &
                 Shift(Line(30,250),10,170);
          man := Shift(torso & head & arms & legs,40,180);
          PutFigure(man);
          torso:= Shift(Line(50, 0),25,180) &
                  Shift(Line(50,60),25,180) &
                  Shift(Line(50,120),25, 0) ;
          head:= Shift(Rotate(Scale(torso,40),180),43,90);
          woman:=Shift(torso & head & arms & legs,40,  0);
          PutFigure(woman);
          couple := man & woman;
          PutFigure(couple & Scale(couple,130) & Scale(couple,160));
end        Marriage;
```

ADA program: Marriage

If we take the result of this program as the data for the program

```
with       Drawing
procedure Market    is
      use Drawing; -- results on plotter
          man,woman,family,trio: FIGURE;
begin  --  again we have cheated slightly
          man    := Scale(GetFigure(),40);
          woman  := Scale(GetFigure(),40);
          family := Scale(GetFigure(),40);
          trio := man & woman & Shift(woman,16,180);
          PutFigure( Shift(family,82,45) & Shift(trio,82,315  ) &
                     Shift(trio,50,90)   & Shift(family,50,270) &
                     Shift(family,82,135)& Shift(man&woman,82,225) );
end        Market;
```

ADA program: Market

we get the drawing

Market drawing

Notice that you can forget the environment Tortoise when you have the environment Drawing. Check that the program

```
with        Drawing
procedure Street   is
       use Drawing;-- results on plotter
           function Box(m,x,y:INTEGER) return FIGURE is
           begin
                     return(Shift(Shift
                           (Line(m,0) & Shift(Line(m,0),m,90) &
                            Line(m,90)& Shift(Line(m,90),m,0)  )
                           ,x,0),y,0);
           end      Box;
           function House(m:INTEGER) return FIGURE is
           begin
                     return(Shift(Box(12*m,  0,   0)  &
                                  Box( 2*m,5*m,12*m)  &
                                  Box( 4*m,4*m,   0)  &
                                  Box( 3*m,2*m,  7*m) &
                                  Box( 3*m,7*m,  7*m)  ,
                                  25*(7-m),20)));
           end      House;
  begin   -- Street
           PutFigure(house(7) & house(6) & house(5) & house(4) &
                     house(3) & house(2) & house(1) );
  end      Street;
```

ADA program: Street

gives a picture you have seen before

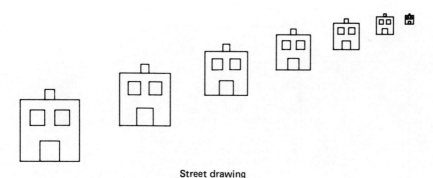

Street drawing

We could have written the module interface as

```
package Drawings is
        type FIGURE is private;
        procedure Line(length,direction:in INTEGER;f:out FIGURE);
        procedure Arc(length,direction,angle:in INTEGER    ;f:out FIGURE);
        function  "&"(left,right:in FIGURE)         return FIGURE;
        procedure Shift (f:in out FIGURE;length,direction:in INTEGER);
        procedure Scale (f:in out FIGURE;procent:in INTEGER);
        procedure Rotate(f:in out FIGURE;  angle:in INTEGER);
        procedure PutFigure(f:in  FIGURE);
        procedure GetFigure(f:out FIGURE);
private -- FIGURE is specified later
end     Drawings;
```

Drawings package interface

You can judge the effects of style by comparing the three Drawing programs with

```
with       Drawings
procedure  Marriage is
      use  Drawings; -- results on plotter
           man,woman,couple,torso,head,f1,f2,f3,
           rightarm,leftarm,rightleg,leftleg:FIGURE;
begin
           Arc(torso,40,90,360);
           Arc(head,15,90,360);Shift(head,40,90);
           Line(rightarm,20,45);Shift(rightarm,28,45);
           Line(leftarm ,20,135);Shift(leftarm,28,135);
           Line(rightleg,30,290);Shift(rightleg,10,10);
           Line(leftleg ,30,250);Shift(leftleg,10,170);
           man := torso & head & rightarm & leftarm
                              & rightleg & leftleg ;
           Shift(man,40,180);          PutFigure(man);
           Line(f1,50, 0);Shift(f1,25,180);
           Line(f2,50,60);Shift(f2,25,180);
           Line(f3,50,120);Shift(f3,25, 0);
           torso:= f1 & f2 & f3; f1:=head;
           Scale(head,40);Rotate(head,180);Shift(head,43,90);
           woman:=torso & head & rightarm & leftarm
                              & rightleg &leftleg
           PutFigure(woman);
           couple := man & woman;
           f2:=couple;Scale(f2,130);f3:=couple;Scale(f3,160);
           PutFigure(couple & f2 & f3 );
      end  Marriage;
```

Revised ADA program: Marriage

```
with       Drawings
procedure  Market   is
      use  Drawings; -- results on plotter
           man,woman,family,trio,f1,f2,f3,f4,f5,f6: FIGURE;
begin
           GetFigure(man);Scale(man,40);
           GetFigure(woman);Scale(woman,40);
           GetFigure(family);Scale(family,40);
           trio:=woman;Shift(trio,16,180);
           trio   := man & woman & trio;
           f1:= family;Shift(f1,82,45);
           f2:= trio;   Shift(f2,82,315);
           f3:= trio;   Shift(f3,50,90);
           f4:= family;Shift(f4,50,270);
           f5:= family;Shift(f5,82,135);
           f6:= man & woman;Shift(f6,82,225);
           PutFigure( f1 & f2 & f3 & f4 & f5 & f6);
      end  Market;
```

Revised ADA program: Market

```
with        Drawings
procedure Street  is
      use Drawings;-- results on plotter
          f:FIGURE;
          function Box(m,x,y:INTEGER) return FIGURE is
                    n,s,e,w:FIGURE;
          begin
                    Line(e,m,0);Line(w,m,0);Shift(w,m,90);
                    Line(n,m,0);Line(s,m,90);Shift(s,m,0);
                    f:= e & n & s & w;
                    Shift(f,x,0);Shift(f,y,90);
                    return f;
          end       Box;
          function House(m:INTEGER) return FIGURE is
          begin
                    f := (Box(12*m,  0,    0)  &
                          Box( 2*m,5*m,12*m)  &
                          Box( 4*m,4*m,    0)  &
                          Box( 3*m,2*m, 7*m)  &
                          Box( 3*m,7*m, 7*m)  );
                    Shift(f,25*(7-m),20);
                    return f;
          end       House;
begin   -- Street
          PutFigure(house(7) & house(6) & house(5) & house(4) &
                              house(3) & house(2) & house(1) );
end       Street;
```

Revised ADA program: Street

Exercise

Make programs to draw some of: tangrams, your name in various sizes, the ground
plan of the Hagia Sophia mosque in Istanbul, a hat.

108

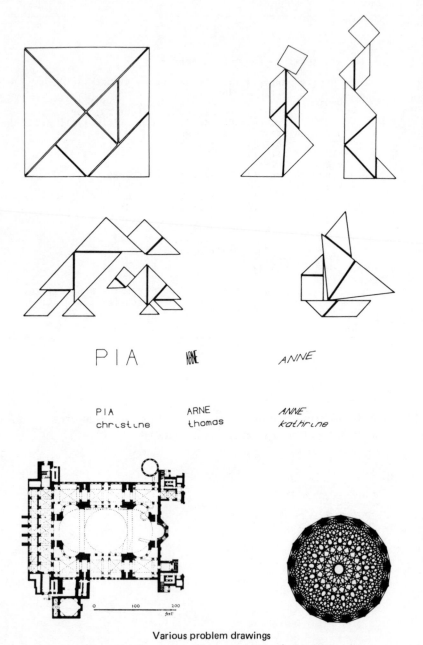

Various problem drawings

(Bottom left drawing reproduced by permission of Prestel-Verlag)

Exercise

If you used the environment Drawing in your solution of the last exercise, give the Drawings solution also; if you used Drawings, give the Drawing solution also.

Exercise

You probably like some features of our two figure drawing environments and dislike others. Design a figure drawing environment that would be ideal for you personally. When you have read this book, return to this exercise and see if increased familiarity with ADA has changed your ideals.

4.3 The environment Syntax and the Jackson method

For our second realistic example of environment design we consider the problem of drawing the structure, syntax and network diagrams in this book.

STATEMENT IS

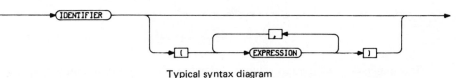

Typical syntax diagram

Such a diagram is a value of the concept RULE which we shall introduce in our environment Syntax. If we obey the environment primitive PutRule(r) in a program which sends its results to the plotter, then the value of r will be drawn. As we want to keep rules in the computer from one program run to another, our environment will also have a primitive GetRule(r) which fetches a rule from its data and assigns the rule to the variable r.

The next step in the design of the environment Syntax is to decide on the primitives for drawing basic components of structure, syntax and network diagrams. If we decide on rectangles, we can have a primitive Rectangle specified by the test data set

RECTANGLE(DATALINE()) IS

Rectangle test data

If we decide on oblongs, we can have a primitive Oblong specified by the test data set

OBLONG(DATALINE()) IS

Oblong test data

As we want more than one oblong or rectangle in our diagrams, our environment provides four primitives—'&', '/', Option, Repeat—with the specification

OBLONG (DATALINE ( )) & RECTANGLE (DATALINE ( )) IS

OBLONG (DATALINE ( )) / RECTANGLE (DATALINE ( )) IS

OPTION (OBLONG (DATALINE ( ))) IS

REPEAT (OBLONG (DATALINE ( )) , RECTANGLE (DATALINE ( )))

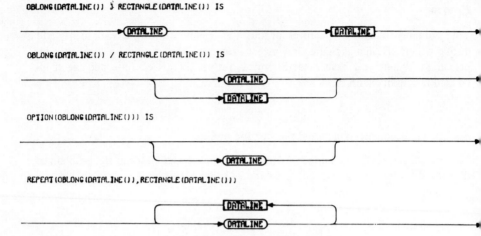

'&', '/', Option and Repeat test data

The user's view of the environment Syntax is given by the primitive specifications and the module interface

```
with      LineManipulator
package   Syntax is
    use   LineManipulator; -- but results on plotter
          type RULE is private;
          function Oblong    (name:LINE) return RULE;
          function Rectangle(name:LINE) return RULE;
          function Option    (r:RULE)    return RULE;
          function "&"(left,right:RULE) return RULE;
          function "/"(left,right:RULE) return RULE;
          function GetRule   (name:LINE) return RULE;
          -- checks that name is title of returned rule
          function Repeat(left,right:RULE) return RULE;
          procedure PutRule (name:LINE;r:RULE) ;
          -- puts name as title of the stored rule
    private -- RULE is specified later
    end       Syntax;
```

Syntax package interface

If we want the first diagram in this section, we might obey the program

```
with      Syntax
procedure StatementDiagram is
    use   Syntax; -- results on plotter
          a,b,c:RULE;
begin
          a := Repeat(Oblong("EXPRESSION"),Rectangle(","));
          b := Rectangle("(") & a & Rectangle(")");
          c := Oblong("IDENTIFIER") & Option(b);
          PutRule("STATEMENT IS",c);
    end       StatementDiagram;
```

ADA program: Typical Syntax Diagram

We should not leave this section on syntax diagrams without mentioning a method for constructing programs that has become popular in industry: the Jackson method. The underlying slogan is 'the structure of a program should reflect the structure of its data'; the method is (1) invent a set of syntax diagrams that describes the possible data for the program, (2) for each syntax diagram devise two algorithms: Before Action to be obeyed when the data 'enters' the diagram. After Action to be obeyed when the data 'leaves' the diagram. We can illustrate this design method by giving an algorithm for summing a sequence of numbers. The possible data are given by the two syntax diagrams

SEQUENCE IS

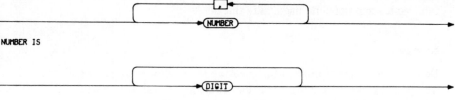

NUMBER IS

Syntax diagrams for data

The two diagrams are given Before Action and After Action algorithms

BEFORE SEQUENCE IS

| SUM:=0; | | DOUBLE SUM:=0 |

AFTER SEQUENCE IS

| PUT SUM AND DOUBLE SUM |

BEFORE NUMBER IS

| DO NOTHING |

AFTER NUMBER IS

| SUM:=SUM+NUMBER | | DOUBLE SUM:=DOUBLE SUM + SUM |

Before Action and After Action structure diagrams

Suppose we combine these algorithms into a program and we supply the data: 11, 73, 512, 1384. The Before Action algorithm for the first syntax diagram gives

sum := 0   ; double sum := 0;

the After Action algorithm for the second syntax diagram gives

sum := 11   ; double sum := 11;
sum := 84   ; double sum := 95;
sum := 596   ; double sum := 691;
sum := 1980   : double sum := 2671;

and the After Action algorithm for the first syntax diagram gives

Sum is 1980
Double Sum is 2671

This problem is unfair to the Jackson method because it is too simple; you can see the method applied to more realistic examples in the book: M. Jackson, *Principles of Program Design,* Academic Press 1975.

Exercise

The syntax diagrams in this book were produced by a more elaborate version of our environment Syntax. Write Syntax programs for some of the diagrams in the book. Suggest new primitives for the environment Syntax so that it can generate all structure, syntax and network diagrams in this book.

Exercise

Design an interactive program for checking the environment Syntax. Compare the way your checker requires syntax rules to be input with the way rules are given in the official ADA definition.

Exercise

The Jackson method can be described by the pipeline

Action Collection < = Parser

In our sum-doublesum problem the Parser might convert ,11, 73, 512, 1384.' to '[(11)(73)(512)(1384)]' where [and] indicate Before Action and After Action for the sequence syntax diagram and ( ) indicate Before Action and After Action for the number syntax diagram. Write an ADA program for the algorithm Action Collection specified by the decision table

Symbol	[	]	(	)	0..9	others
Effect	Start	Finish	a:=0	Accumulate	a:=a*10+symbol	ErrorMessage

Action Collection

4.4 The environments Text Manipulator and Book Production

For our third realistic example of environment design we consider the problem of writing and correcting programs, letters and other documents without pictures. In our environment Text Manipulator we will have a concept TEXT whose values are sequences of lines. We will also have a primitive Put Text for adding a sequence of lines to result and a primitive Get Text for fetching a sequence of lines from data. All the other primitives in our environment will assume texts are sets of numbered lines like

    5: We can think of
    10: values of TEXT
    20: sequences of numbered lines.

We want the lines to be numbered when we are making corrections but we do not mind if Put Text removes numbers and Get Text invents a new numbering.

Our environment Text Manipulator will provide a primitive Correct for making corrections of three kinds (1) inserting a new line (2) replacing an old line (3) removing one or more old lines. The specification for Correct might be the test data set

```
text   T              5: We can think of
                     10: values of TEXT
                     20: sequences of numbered lines.

keyboard line L      25  If we want to!

Correct(T,L)          5: We can think of
                     10: values of TEXT
                     20: sequences of numbered lines.
                     25: If we want to!

keyboard line L      10  values of TEXT as

Correct(T,L)          5: We can think of
                     10: values of TEXT as
                     20: sequences of numbered lines.


keyboard line L      5-15

Correct(T,L)         20: sequences of numbered lines.
```

Correct test data

Like the primitive New Line in the environment Line Manipulator, the primitive Correct will always take information from the display keyboard, even if data ≠ keyboard. An algorithm for correcting a text might be

MANY CORRECTIONS IS

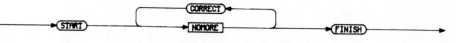

Many corrections structure diagram

but we would like the environment to help us out of the loop by providing an exception: No More Input. Assuring that we can type something on the display keyboard to make Correct raise this exception we can convert our algorithm into an ADA program

```
with       TextManipulator
procedure EditData is
       use TextManipulator;
       t  :TEXT;
begin
          t := GetText();
          loop
               Correct(t);
          end  loop;
exception
          when NoMoreInput => PutText(t);
end        EditData;
```

ADA program: Edit Data

114

Note that this program can be used instead of the computer primitive Edit(data, result) when data ≠ empty. Thinking about Edit(empty,result) suggests the program

```
with        TextManipulator
procedure EditEmpty is
        use TextManipulator;
        t  :TEXT;
begin
        t := Empty();
        loop
                Correct(t);
        end loop;
exception
        when NoMoreInput => PutText(t);
end        EditEmpty;
```

ADA program: Edit Empty

Our environment provides the primitive Empty Text—giving a text consisting of *no* lines—for two reasons (1) so we can write the above program (2) so we have a way of generating a text (remember the problem—which came first the hen or the egg). The last primitive in our environment is Renumber with the specification

```
        data D                          Renumber(D)
 1: TRANSLATE(program,code)       2: TRANSLATE(program,code)
 2: followed by                   4: followed by
 4: DO (code,empty,results)       6: DO (code,empty,results)
 5: is a clumsy way               8: is a clumsy way
15: of saying                    10: of saying
20: EDIT(empty,results)          12: EDIT(empty,results)
```

Renumber test data

We can use this primitive when we want to insert a new line between two old lines and there is no room.

The interface for Text Manipulator is

```
with      LineManipulator
package TextManipulator is
        use LineManipulator;-- results on printer
        type TEXT is private;
        NoMoreInput:EXCEPTION;
        function  Empty        return TEXT;
        function  GetText      return TEXT;
        procedure PutText (t:TEXT);
        procedure Correct (t:in out TEXT);
        procedure Renumber(t:in out TEXT);
private -- TEXT is specified later
end      TextManipulator
```

Text Manipulator package interface

The environment Text Manipulator underlies the environment Book Production in Chapter 2. There you saw the specification of the primitives—Normal, Serif, Gothic, Greek, Russian, Underline—and you met a program Justify that is suitable for fonts whose characters are all of the same size. The specification for the Book Production primitive Justify might be

ιιιαααιιι
ωωωμμμωωω

becomes

ι ι ι α α α ι ι ι
ωωωμμμωωω

Justification of varied fonts

The other Book Production primitives are used to fetch data and save results:

Put Page — saves results of a Book Production program
Get Page — fetches information saved by Put Page
Figure To Page — fetches information saved by Put Figure in a Drawing or Drawings program
Rule To Page — fetches information saved by Put Rule in a Syntax Diagram
Text To Page — fetches information saved by Put Text in a Text Manipulator program

The user's view of the environment Book Production is given by the primitive specification and the interface

```
package BookProduction is
  -- results on photosetter
      procedure Normal;
      procedure Serif;
      procedure Gothic;
      procedure Greek;
      procedure Russian;
      procedure Underline;
      procedure Justify;
      procedure FetchLine;
      procedure FigureToPage;
      procedure RuleToPage;
      procedure TextToPage;
      procedure PutPage;
      procedure GetPage;
end     BookProduction;
```

Book Production package interface

Were you surprised by the fact that none of the primitives had parameters and the interface introduced no variable or type. You will be less surprised when you have seen the Book Production program that generates the specification of the primitive Justify:

116

```
with       BookProduction
procedure  Page1 is
     use BookProduction; -- results on photosetter
begin
     FetchLine;   -- "iiiaaaiii" from keyboard
     Greek;
     FetchLine;   -- "wwwmmmwww" from keyboard
     Greek;
     FetchLine;   -- " becomes " from keyboard
     PutPage;   -- gives the first three lines
     FetchLine;   -- "iiiaaaiii" from keyboard
     Greek;
     FetchLine;   -- "wwwmmmwww" from keyboard
     Greek;
     Justify;   -- combines last two lines
     PutPage;
end        Page1;
```

ADA program: Page 1

So that you can see some of the other primitives in action we give a program that generates a page with a syntax diagram, a drawing and two titles

```
with       BookProduction
procedure  Page2 is
     use BookProduction; -- results on photosetter
begin
     RuleToPage;   -- gets syntax diagram from data
     FetchLine;      -- gets title from keyboard
     FigureToPage;-- gets drawing from data
     FetchLine;      -- gets title from keyboard
     PutPage;
end        Page2;
```

ADA program: Page 2

Exercise

Design an interactive program for checking the environment Book Production. What information do you have to give to your program to make it generate the different versions of the alphabet in chapter 2?

4.5 Data bases

Many millions have been invested in data base systems. A data base system allows one to keep structured information in the computer, to access this information easily and to restructure it. Depending on the way the information is structured a data base system is said to be hierarchic, network or relational; in this section we shall describe two environments for a relational data base system.

A relational data base system assumes that information is kept as a collection of relations; a data base for the author of a book might well have

picture relation	: for remembering pictures
text-relation	: for remembering texts
reference-definition	: for remembering definitions
reference-bibliography	: for remembering references to other books
reference-others	: for remembering other references

A relation consists of a title and zero, one or more elements; the title of a relation consists of a finite number of different names, and all elements in a relation must be different. The current state of an author's book might be given by

```
picture_relation:SECTION   LINE PICTURE
              4.2            73 LineData
              4.2           117 ArcData
              4.2           226 Street
              4.3             5 Integer

text_relation    :SECTION HEADING    STATUS
              4.1         Interfaces rewritten
              4.2         Drawing    written
              4.3         Figure     concept
              4.3         Syntax     written
              4.5         Databases  progressing

reference_definitions :REFERENCE SECTION LINE
                       interface  4.1        2
                       body       4.1       35
                       text       4.4       14

reference_bibliography:REFERENCE SECTION LINE
                    -- title but  no  elements

reference_others      :REFERENCE SECTION LINE
                       interface  4.1        2
                       body       4.1       35
                       justify    4.4      213
```
Book data base status

Our environments will provide three primitives for combining two relations with the same title into another relation with the same title

* keeps elements in both relations
+ keeps elements in the first relation, the second relation or both
− keeps elements in the first relation but not the second relation

If we apply these primitives to the relations reference-definition and reference-others, we get the three relations

```
reference_definitions * reference_others
                       :REFERENCE SECTION LINE
                        interface  4.1       2
                        body       4.1      35

reference_definitions + reference_others
                       :REFERENCE SECTION LINE
                        interface  4.1       2
                        body       4.1      35
                        text       4.4      14
                        justify    4.4     213

reference_definitions - reference_others
                       :REFERENCE SECTION LINE
                        text       4.4      14
```
'*', '+', '−' test data

Our environments will provide one and only one primitive for combining two relations with different titles. If we apply this primitive Join to picture-relation and text-relation, we get

:SECTION	LINE	PICTURE	HEADING	STATUS
4.2	73	LineData	Drawing	written
4.2	117	ArcData	Drawing	written
4.2	226	Street	Drawing	written
4.3	5	Integer	Syntax	written
4.3	5	Integer	Figure	concept

Join test data

The precise definition of the result of joining two relations varies from one relational data base system to another. Our precise definition is (1) a relation B can be joined to a relation C if there are distinct names $a, b_1, \ldots, b_m, c_1, \ldots, c_n$ such that the title of B is $(a, b_1, \ldots, b_m)$ and the title of C is $(a, c_1, \ldots, c_n)$ (2) the result of joining a relation B with title $(a, b_1, \ldots, b_m)$ to a relation C with title $(a, c_1 \ldots, c_n)$ is a relation with title $(a, b_1, \ldots, b_m, c_1, \ldots, c_n)$ (3) there is an element $(x_0, x_1, \ldots, x_{m+n})$ in the result of joining B and C if and only if there is an element $(x_0, x_1, \ldots, x_m)$ in B and and element $(x_0, x_{m+1}, \ldots, x_{m+n})$ in C. Notice that we can only join relations when their titles have the same first name. This restriction is not as serious as it might appear, because our environments will provide a primitive Project for rearranging relations. If we project text-relation according to the title (SECTION HEADING) and we project reference-definition according to the title (SECTION REFERENCE) we get

:SECTION	HEADING
4.1	Interfaces
4.2	Drawing
4.3	Syntax
4.3	Figure
4.5	Databases

:SECTION	REFERENCE
4.1	interface
4.1	body
4.4	text

Project test data

Because the titles of these two relations have the same first name, we can join them to get

:SECTION	HEADING	REFERENCE
4.1	Interfaces	interface
4.1	Interfaces	body

Another test of Join

The precise definition of Project is (1) a relation R can be projected according to the title $(t_1, t_2, \ldots, t_m)$ if the names $t_1, \ldots, t_m$ occur in the title of R, (2) the result of projecting a relation according to a title $(t_1, t_2, \ldots, t_m)$ is a relation with title $(t_1, t_2, \ldots, t_m)$, (3) there is an element $(x_1, x_2, \ldots, x_m)$ in the result of projecting a relation R according to the title $(t_1, t_2, \ldots, t_m)$ if and only if there is an element $(r_1, \ldots, r_n)$ in R such that: if x_i and r_j have the same title name, then $x_I = r_j$.

Our data base environments must provide some ways of creating relations. The primitive Create will read an element $(x_1, \ldots, x_m)$ from the display keyboard and return a relation with $(x_1, \ldots, x_m)$ as its one and only element. The design decision to provide elements from a display keyboard also motivates the primitives for extracting information from a data base. Our environment will have a primitive Same for finding those elements in a relation whose first component is the same as a given value. If we apply Same to picture-relation and give the value '4.3', it will give the one-element relation

:SECTION	LINE	PICTURE
4.3	5	Integer

Same test data

if we apply Same to text-relation and give the value '4.3', it will give the two-element relation

:SECTION	HEADING	STATUS
4.3	Syntax	written
4.3	Figure	concept

Same test data

The other two primitives for extracting information are Before—for extracting elements whose first components are earlier than a given value—and After—for extracting elements whose first components are later than a given value. If we apply Before and After to picture-relation and give the value '4.3', we get the relations

:SECTION	HEADING	STATUS
4.1	Interfaces	rewritten
4.2	Drawing	written

:SECTION	HEADING	STATUS
4.5	Databases	progressing

Before and After test data

The user's view of the environment Private Data Base is given by the primitive specifications and the interface

```
with      TableManipulator -- given later
package PrivateDataBase is
   use TableManipulator;
        type RELATION is limited private;
        function "*" (first,second:RELATION)         return RELATION;
        function "+" (first,second:RELATION)         return RELATION;
        function "-" (first,second:RELATION)         return RELATION;
        function Join(first,second:RELATION)         return RELATION;
        function Project(r:RELATION;title:LINE) return RELATION;
        function Create (title:LINE)                 return RELATION;
             -- reads an element from keyboard
        function Same  (r:RELATION)                   return RELATION;
             -- reads a key from keyboard
        function Before(r:RELATION)                   return RELATION;
             -- reads a key from keyboard
        function After (r:RELATION)                   return RELATION;
             -- reads a key from keyboard
        function GetRelation                          return RELATION;
        procedure PutRelation(r:RELATION);
        BadParameters:EXCEPTION;
   private type RELATION is given later;
   end      PrivateDataBase;
```

Private Data Base package interface

Suppose our author wants to record a day's work in his data base by updating picture-relation and text-relation. He can do this by running the program

```
with      PrivateDataBase
procedure TodaysChanges  is
   use PrivateDataBase;
   pr,tr:LINE;c:CHARACTER;
   pictures,texts:RELATION;
begin
        pictures:=GetRelation();pr:=DataLine();
        texts   :=GetRelation();tr:=DataLine();
        loop
           Get(c);
           case c is
              when 'a'    => pictures:=pictures + Create(pr);
              when 'b'    => pictures:=pictures - Create(pr);
              when 'c'    => texts    :=texts    + Create(tr);
              when 'd'    => texts    :=texts    - Create(tr);
              when others => exit;
           end  case;
        end  loop;
        PutRelation(pictures);
        PutRelation(texts);
   end      TodaysChanges;
```

ADA program: Todays Changes

The reason why so much money has been invested in data base systems is that they provide a convenient way of retrieving information. Ideally we should have

— well-defined limits on what information can and cannot be revealed to any given person;
— the possibility of phrasing questions in ordinary natural language;

but the data base systems used in practice are such that

— with a little imagination any programmer can get almost all information because protection systems are feeble;
— questions have to be formulated in a very artificial language.

If our author wants to know

— what pictures are in section 5
— which pictures occur in the same section as a definition

he can run the programs

```
with        PrivateDataBase
procedure Question1 is
        use PrivateDataBase;
        pictures:RELATION;
begin
        pictures:=GetRelation();        -- picture_relation
        pictures:=Project(pictures,"SECTION PICTURE");
        PutRelation( Same(pictures)); --"4.5"from keyboard
end         Question1;
```

```
with        PrivateDataBase
procedure Question2 is
        use PrivateDataBase;
        pictures,definitions:RELATION;
begin
        pictures := GetRelation(); -- picture_relation
        definitions:=GetRelation(); -- reference_definition
        PutRelation(
                Join( Project(pictures,   "SECTION PICTURE"  )
                    , Project(definitions,"SECTION REFERENCE")));
end         Question2;
```

ADA program: Question 1, Question 2

These ADA programs are very clumsy but not as bad as they would have been if our author's questions had to be reformulated in the Procrustean beds provided by some commercial data base systems.

Data base experts tend to think of computers running different programs on the same data, while other programmers tend to think of computers running the same program on different data. In a picture with tortoises for programs and boxes for data

122

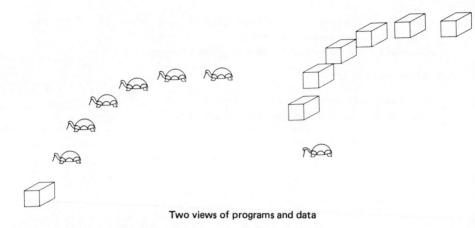

Two views of programs and data

The most impressive data base system known to the author is that developed in three man years for the Danish ship builders Burmeister & Wain. The main data base relations are so large that they are stored on several magnetic tapes. A typical tape is generated once a month from a tape provided by the Lloyds insurance company in London, and it contains 200 component tuples for each of the 80,000 ships currently afloat. Ordinary users of this data base system, MIAS, cannot change the main data base relations but they have 1) high speed utilities to copy these relations to relations in their *shadow data bases,* 2) relational operations like those in our module Private Data Base for manipulations on the 20 component 10,000 tuple relations in a shadow data base, 3) powerful report generating facilities for printing relations in a form fit for humans.

The MIAS solution to the Procrustean bed problem is to prepare question schemas for the common user questions, and leave the programming of uncommon questions to a couple of specialists who know about operations like join. The specialists handle about half of the user questions, and question schemas handle the rest.

There is one important difference between MIAS and the data base systems used by insurance offices, booking agencies and the like: each user has her own shadow data base. How can we avoid this restriction, how can we modify our 'single user' environment Private Data Base so many users can explore a data base at the same time? One answer is that we can replace Private Data Base by an environment that uses ADA's task concept:

```
with      TableManipulator
task      PublicDataBase    is
    use TableManipulator;
       type  RELATION is limited private;
       entry Product(f,s:RELATION;          result:out RELATION);
       entry Sum     (f,s:RELATION;          result:out RELATION);
       entry Differ (f,s:RELATION;          result:out RELATION);
       entry Join    (f,s:RELATION;          result:out RELATION);
       entry Project(r:RELATION;t:LINE;  result:out RELATION);
       entry Create  (title:LINE;            result:out RELATION);
       entry Same    (r:RELATION;key:LINE;result:out RELATION);
       entry Before (r:RELATION;key:LINE;result:out RELATION);
       entry After   (r:RELATION;key:LINE;result:out RELATION);
       entry PutRelation(r:in  RELATION;user:LINE);
       entry GetRelation(r:out RELATION;user:LINE);
private type  RELATION is -- specified later;
end       PublicDataBase;
```

Public Data Base task interface

Here we notice (1) procedures and functions have become task entries (2) *return* cannot be used in task interfaces (3) PutRelation and GetRelation require user identification. There are two reasons why the concept RELATION has been introduced in the LIMITED way and not the PRIVATE way; the first is that we want to prevent two users assigning to the same relation at the same time; the second is that this gives the environment the possibility of keeping a 'logbook' on who modified which relation when. If Public Data Base is being used by a real-life organization like a bank, such a logbook can uncover dishonest users. The three main defenses against computer fraud are

—— keep a logbook

—— invent a system of passwords to protect against damage by unauthorized users

—— encode critical information (remember the crytography exercise at the end of Chapter 1).

Programs that are used in sensitive organizations are not documented properly unless there is a section on: protection against deliberate and accidental misuse.

Exercise

Rewrite the programs Todays Changes, Question 1, Question 2 so that they work in the environment Public Data Base. How can you redesign the environment Private Data Base so that all programs that work for that environment also work for the environment Public Data Base and vice-versa? Remember that users neither know nor care if they are calling an *entry* or *procedure*.

4.6 Environment design

An ADA environment is a collection of computer algorithms for solving a variety of problems, in the same way that an ADA program is a computer algorithm for solving a particular problem. However designing an environment requires more care than defining a program because a useful environment must be coherent, its algorithms must be interrelated. The environments in this chapter are coherent because we have explained how they can be used without giving details of the algorithms they provide.

If an environment is coherent, it is easy to design a program for testing its behaviour. A test program for the environment Tortoise could be an ADA version of the decision table

Character	d	u	m	t	c	s	e	n	w	others
Action	down	up	move num	turn num	turnto num	south	east	north	west	
	again	again	again	again	again	again	again	again	again	

This decision table can be converted into the program

```
with        Tortoise,LineManipulator
procedure TortoiseTest is
        use Tortoise,LineManipulator; -- results on plotter
            control:CHARACTER;number:INTEGER;
begin
        loop
                Get (control);
                case control is
                        when    'd' => Down;
                        when    'u' => Up;
                        when    'm' => Get(number);Move(number);
                        when    't' => Get(number);Turn(number);
                        when    'c' => Get(number);TurnTo(number);
                        when    's' => South;
                        when    'e' => East;
                        when    'n' => North;
                        when    'w' => West;
                        when    others=> exit;
                end case;
        end loop;
end     TortoiseTest;
```

ADA program: Tortoise Test

A good test program should be more robust than this; incorrect data should not cause it to collapse, control data should be distinguished from parameter data, there should be an extended form of control data. Our test program should accept data in the form

. east
. north
. turn (90)

as well as the abbreviated form: e n t 90. The Tortoise test program should have been

```
with       Tortoise,LineManipulator
procedure  TortoiseTest is
      use  Tortoise,LineManipulator;  -- results on plotter
           l:LINE;number:INTEGER;anotherT:BOOLEAN;
begin
           loop
                case LineToCharacter() is
                     when     'd' => Down;
                     when     'u' => Up;
                     when     'm' => Get(number);Move(number);
                     when     't' => anotherT:=FALSE;
                                     loop
                                          case NextCharacter() is
                                               when   'T' =>
                                                    anotherT:=TRUE;
                                               when '('!LF  => exit;
                                          end case;
                                     end loop;
                                     Get(number);
                                     if     anotherT
                                     then   TurnTo(number);
                                     else   Turn(number);
                                     end    if;
                     when     'c' => Get(number);TurnTo(number);
                     when     's' => South;
                     when     'e' => East;
                     when     'n' => North;
                     when     'w' => West;
                     when     'z' => exit;
                     when  others=> Put("Error:Type z for exit");
                end case;
           end loop;
      end  TortoiseTest;
```

Revised ADA program: Tortoise Test

Two good reasons for designing an environment test program are

— so users can become familiar with the environment algorithms
— so users can show the environment designer what they think are errors, without the designer having to examine obscure user programs.

For the same reasons a test program should be designed before the environment algorithms when a group of programmers are cooperating in the development of an environment.

Most of the money spent on solving problems by computers goes to people tracking down unobvious, stubborn errors, in spite of the fact that most programming errors are easy to find and correct. This is the socalled software crisis and the reason why ADA has been created. However you will still make unobvious errors in your programs and you should know three techniques for eliminating them—test data design, desk checking and tracing. We shall illustrate these techniques on a program that does not solve the problem:

Given non-zero numbers $a_1 \, a_2, \ldots, a_n$ followed by zero,
find the sum $a_1 + a_2 + \ldots + a_n$
and the double sum $a_1 + (a_1 + a_2) + \ldots + (a_1 + a_2 + \ldots + a_n)$

```
with        Text_IO
procedure Accumulate is
        use Text_IO;
        number,sum,double_sum:INTEGER;
begin
        sum := 0; double_sum := 0;
        loop
                Get(number);
                exit when    number=0;
                double_sum:=double_sum+sum;
                sum := sum + number;
        end    loop;
        Put(sum);Put(double_sum);
    end        Accumulate;
```

ADA program: Accumulate

The design of test data for such a program is an art learnt by experience but there are two general principles: choose some data where the desired result is obvious, choose other data for unusual situations that might arise. A reasonable set of test data for our sum-doublesum problem might be

data	0	1,0	1,1,0	1,2,3,0	3,2,1,0
sum	0	1	2	6	6
doublesum	0	1	3	10	14

If we give this test data to our program, it will give incorrect values of doublesum. To find out why, we can make a desk check—we make a table with a row for each variable and we simulate each step of the program

a	1	2	3	0	
sum	0	1	3	6	put
doublesum	0		1	4	put

The program will incorrectly give 4 as the double sum, but the process of making the table probably showed that the program is incorrect because doublesum gets a new value before sum gets a new value. It is easy to make a mistake, when simulating each step in a program, so you should resort to our next debugging technique—tracing—if a desk check does not reveal the cause of an error. One traces a program by making it print appropriate intermediate results—if we insert

Put(number) ; Put(sum) ; Put(doublesum)

after Get(number) in our program, we get so much of the desk check table that the program error is easy to see. More generally: if we make a program print parameter values before and after every call of an environment procedure, function or entry, we usually get enough information to decide whether an error is in the program or its environment.

What should we do when we cannot find an error in an incorrect program? First we should check that the test data has exercised every line in the program; if it has not, we should devise more test data. When the program behaves correctly for test data that exercises every line in the program, we can complete the documentation of our problem solution by giving both the program and the test data. As this section is supposed to be about environment design, we should close by saying how to complete the documentation of an environment when it has been tested adequately. Because environments may be redesigned from time to time— e.g. to improve efficiency—their documentation should contain a *guarantee*, a promise that user programs covered by the guarantee will always behave correctly no matter what happens to the environment. Ideally the guarantee in the documentation should be precise; in practice natural language is ambiguous and formal language is foreign to the expected users. Unfortunately the ADA interface and environment test program give absolutely no information about the behaviour of the environment algorithms, and the test data can only give this information for a few particular cases.

The ADA programs in this book are a guarantee for the 'syntax checker' program the author has used, but they may still be incorrect because errors may have crept in when the syntax rules were transferred from the language definition to the syntax checker.

Exercise

Write test programs for some of the environments in this chapter. Find suitable test data for these environments and write the environment guarantee.

Exercise

An environment can react in many ways to inappropriate data. Think of situations where you would like an environment

- to send you a message
- to ignore your data completely
- to handle your data until it discovers an error, and ignore the rest
- to make an intelligent guess when it discovers an error
- to raise an exception.

Exercise

In the Tortoise environment TurnTo(90) always gives the value 90 to the variable direction, while the value given by Turn (90) may be anything. The procedure TurnTo is 'honest' in a way that the procedure Turn is not. Make this notion of honesty precise. Which of the environment procedures you have seen are honest? Should the documentation of an environment mention honesty?

Chapter 5

Types

If we are to be sure that a computer program solves the problem it is supposed to solve, then we would like the computer to check that the program is consistent. In this chapter we describe how the computer can check the consistency of a program in ADA because every variable and parameter in the program must have a type.

5.1 Variables and expressions

In section 2.4 we distinguished between algorithms that produced an interesting value and those that did not. Suppose we have an algorithm C which produces a number when given two numbers as arguments. If A and B are algorithms, which produce a number and require no arguments, we can use C to construct many new algorithms.

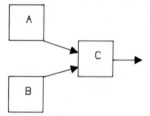

Typical expression structure diagram

In the case when A produces the number 17, B produces the value of the variable b and C is multiplication we get

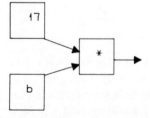

Example expression structure diagram

128

In structure diagrams algorithms that produce values become rectangles, in **ADA** they become *expressions*

C (A,B)
17 * b

You should check that these expressions fit the syntax diagrams

EXPRESSION IS ALSO

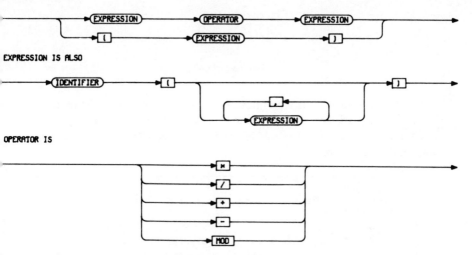

EXPRESSION IS ALSO

OPERATOR IS

Expression syntax diagrams

You should know that the operators in the last of these syntax diagrams are governed by priority rules: '2 + 3 X 4' gives the same value as 2 + (3 X 4) and a different value from (2 + 3) X 4. We do not give the priority rules here, because you can always use parentheses when you do not know or remember the priority rules.

If A is an algorithm that produces a value then obeying the algorithm

Assignment structure diagram

places this value in the variable v. In ADA every variable has a type, and the computer checks that a value belongs to this type before it is placed in the variable. If our variable v has the type INTEGER, then the computer checks that the value produced by our algorithm is a number. To see how variables and parameters get a type, let us look at the two ADA named units

```
procedure  Pap  is
        x   :INTEGER;
begin
            x := 2;
            Put(Pip(x) - Pip(2));
end         Pap;



function  Pip(formal:INTEGER)  return  INTEGER  is
        y   :INTEGER;
begin
            x := formal + x;
            y := x + z;
            return  x+2;
end         Pip;
```

Two named units

Variables in named units are of three kinds; if they come from the environment, they are global variables; if they are introduced in the named unit itself, they are local variables; otherwise they are inherited variables. When we combine our two named units into a program

```
with      Mystery    -- includes Text_IO and variable  z
procedure Pap     is
        use Mystery
        x   :INTEGER;
            function Pip(formal:INTEGER) return INTEGER is
                y   :INTEGER;
            begin
                    x := formal + x;
                    y := x + z;
                    return x+2;
            end         Pip;
begin
                x := 2;z:=2;
                Put(Pip(x)-Pip(2));
end         Pap;
```

ADA program: Pap

we see that Pip has the global variable z, the local variable y and the inherited variable x. Global and inherited variables allow information to flow between the solution of a subproblem and the solution of a large problem, but uncontrolled information flow can be dangerous. You may have noticed that most of the assignment statements in our example have no effect on the number produced by the program. Did you also notice that the number produced by the program depends on whether Pip(x) was obeyed before Pip(2); if Pip(x) is obeyed before Pip(2), Pip(x) will produce 6, Pip(2) will produce 8, and the program will produce −2; if Pip(x) is obeyed after Pip(2), Pip(2) will produce 6, Pip(x) will produce 10 and the program will produce +4. Because such side effects are the source of many programming errors, it is usually wise to protect yourself by using parameters instead of global and inherited variables. Within a procedure parameters behave like local variables; the difference between the two calls of Pip in our example is that Pip(x) obeys the assignment 'formal := x' before obeying the rest of Pip, whereas Pip(2) obeys the assignment 'formal := 2'. Such implicit assignments occur whenever one calls a procedure with parameters. Consider the case when we have the call P(actual) of

procedure P (formal: INTEGER) is . . .

Before doing anything else, the procedure P will obey the implicit assignment 'formal := actual'. We could have emphasized this implicit assignment by writing the declaration

procedure P(formal: in INTEGER) is . . .

We get quite a different implicit assignment from the declaration

procedure P (formal: out INTEGER) is . . .

In this case the call P(actual) would cause the procedure to obey 'actual := formal' after doing everything else. If we have written the declaration

procedure P (formal: in out INTEGER) is . . .

the procedure would have obeyed the two implicit assignments

formal := actual   before doing anything else
actual := formal   after doing everything else

Judicious use of *in, out, in out* parameters makes for robust and intelligible programs.

Intelligible programs usually use values and variables of many types. The program in this section is still meaningful if we replace ': INTEGER' by ': ANOTHERTYPE' and (1) the environment Mystery gives a meaning to ANOTHER TYPE, (2) the expression '2' is a value of ANOTHER TYPE, (3) expressions involving + can have values of type ANOTHER TYPE. Soon we will reveal when the first two requirements are satisfied, here we just mention the fact that (3) is satisfied if and only if the environment Mystery overloads '+' appropriately.

5.2 Constraints and coercions

Programs should deliver error messages rather than incorrect results. In practice we often know a constraint that must be satisfied by the values of the variables, if the program is behaving correctly. Remember the variable horizontal in the environment Tortoise. If it had been introdced with a constraint

horizontal: INTEGER *range* −100 . . . +100

then a program using the environment Tortoise would protest if it was asked to move the plotter pen over the edge of the plotter paper. As an example of a program that checks constraints on values consider

```
with        Text_IO
procedure Checker is
       use Text_IO;
       i   :INTEGER;
       c   :CHARACTER;
       l   :CHARACTER range 'a'..'z';
begin
             l := 'c'; -- check 'c' between 'a' and 'z'
             c := 'l'; -- no check
             l :=  c ; -- check value of c between 'a' and 'z'
             l :=  2 ; -- protest because 2 is an INTEGER
             l := '2'; -- protest because '2' is after 'z'
             i :=  2 ; -- no check
             i := '2'; -- protest because '2' is a CHARACTER
       end      Checker;
```

ADA program: Checker

Notice the difference in meaning between the assignments

c := 1; c := "l"; l := 2; l := "z";

the distinction between the character "l" and the variable l is similar to the distinction between the character "2" and the number 2. Notice also that type errors like l:= 2 and i := "2" are caught by the ADA translator, while constraint errors like l:= "2" raise an exception when the program is obeyed.

Some programs become more readable as well as more robust when value constraints are given by subtypes or derived types. Subtypes are used in the program

```
with        Tortoise
procedure SafeRectangle is
       use Tortoise;
       subtype SAFELENGTH is INTEGER range 0..100;
       width,height:SAFELENGTH;
begin
             Down;
             Get(width);Get(height);
             Move(width); Turn(90);
             Move(height);Turn(90);
             Move(width); Turn(90);
             Move(height);Turn(90);
             Up;
exception when Constraint_Error => null; -- draws nothing
       end      SafeRectangle;
```

ADA program: Safe Rectangle

This program would have been just as robust if the variables had been introduced by

width, height: INTEGER range 0 .. 100;

but it would not have been so readable.

Subtypes give one kind of protection against incorrect assignments; derived types give another. If we assign a derived type to the variable direction by writing

type ANGLE is new INTEGER;
direction: ANGLE;

then the program will protest if it is asked to obey assignment statements like

width := direction;
direction := width;

On the rare occasions when one does want to write assignment statements like these, one can write them as

width := INTEGER (direction);
direction := ANGLE (width);

and the program will *coerce* a value from one type to another.

A convenient way of documenting the types in a program is to give a type hierarchy diagram like:

Type hierarchy diagram

This diagram documents the types in the last program and

```
with        Text-IO
procedure SubtypeTest is
       use Text-IO;
            subtype LETTER is CHARACTER range 'a'..'z';
            subtype HEXLETTER is LETTER range 'a'..'f';
            subtype DIGIT  is CHARACTER range '0'..'9';
        c   :CHARACTER;l:LETTER;h:HEXLETTER;d:DIGIT;
begin
            Get(c);
            l:=c;h:=c;d:=c;
       --   Why is the runtime exception
       --   Constraint-Error always raised?
end         SubtypeTest;
```

ADA program: Subtype Test

As you see from these programs the ADA type checking rules are complicated. We can distinguish between four kinds of assignment statements if we define:

 — a type TB is a subtype of a type TA when we can follow double arrows from TA to TB in the type hierarchy diagram.
 — a type TB is a relative of a type TA when we can follow single or double arrows from TA to TB in the type hierarchy diagram.

The four kinds of assignment statements A := B are:

— no check because the type B is a subtype of the type of A,
— to be checked when the program is obeyed because the type of A is a subtype of the type of B and the types are different,
— protest because the type of A is not a relative of the type of B,
— protest in spite of the fact that the type of A is a relative of the type of B because the assignment is not in the form A := TA (B).

It would be very inconvenient if we could not use such primitive algorithms as Put and Get with relatives of the types INTEGER and CHARACTER. Not only does ADA allow this but it also allows us to use

— the primitives + * − / on relatives of INTEGER;
— the primitives > >= <= < on relatives of INTEGER and CHARACTER;
— the primitives = /= on all types
— the predefined attributes 'FIRST 'LAST on all constrained types.

Exercise

Introduce value constraints in some of the programs in earlier chapters. If you use subtypes and derived types, draw the appropriate type hierarchy diagram.

5.3 Scalars, booleans and constants

Suppose you are writing a program for printing texts in many fonts. You will probably feel a need for a variable to remember which font the photosetter is using at any given moment. It would be unnatural to use numbers or characters as the values of such a variable. Instead you can make a declaration like

current_font: (normal,serif,gothic,greek,russian);

and the computer will check that any value assigned to the variable is one of: normal, serif, gothic, greek, russian. To convince you that user defined value constraints of this kind can make programs readable and robust, we give a program that finds the number of days between two given dates.

```
with        Text_IO
procedure DayCount is
      use Text_IO;
      type MONTH is (january,february,march,april,may,june,july
                      ,august,september,october,november,december);
      m   :MONTH; n,y:INTEGER; d:INTEGER range 1..31;
      month_length:array(MONTH) of INTEGER;
begin
      month_length := (31,28,31,30,31,30,31,31,30,31,30,31);
      d:=4;m:=december;y:=1937;Get(n);
   --  my birthday can be replaced by your birthday
      loop
            if    d < month_length(m)
            then  d:= d+1;              -- d set to next day
            elsif m < december
            then  d:=1;
                  m:= MONTH'succ(m);-- m set to next month
            else  d:=1;m:=january;
                  y:=y+1;              -- y set to next year
            end   if;
            --  adjust for leap year
            if    y mod 4 /=0 then month_length(february):= 28;
            elsif y mod 100=0 then month_length(february):= 28;
            else                    month_length(february):= 29;
            end   if;
      end loop;
      Put(d);Put(MONTH'pos(m));Put(y);NewLine();
   --  'pos converts the value of m to an integer
      end         DayCount;
```

ADA program: Day Count

In this program we have defined the type MONTH by listing its possible values. Such types are called scalar or enumeration types, and they can be used in the same way as INTEGER and CHARACTER to produce subtypes and derived types.

ADA provides a scalar type which is very useful for converting conditional algorithms to programs. In practice we often want to use a variable to remember whether a condition is satisfied or not so ADA provides the scalar type

type BOOLEAN is (FALSE, TRUE)

and allows statements like

above := x * h <= y * v;
if above then EAST else NORTH end if;

The type BOOLEAN is supplied with some useful operators:

p	q	*not* p	p *and* q	p *or* q	p x *or* q
TRUE	TRUE	FALSE	TRUE	TRUE	FALSE
TRUE	FALSE	FALSE	FALSE	TRUE	TRUE
FALSE	TRUE	TRUE	FALSE	TRUE	TRUE
FALSE	FALSE	TRUE	FALSE	FALSE	FALSE

136

You will soon learn to appreciate these operators when you are making a program out of an algorithm whose structure diagram contains conditions and double arrows. You may run into the problem that you want the expression 'p *or* q' to give the value TRUE when the value of p is TRUE and evaluating q would raise an exception. If so, you will welcome the fact that ADA allows you to write 'p or else q' and avoid evaluating q when p is TRUE. Analogously '(x /= 0) and then (1 /x < y)' avoids the exception raised by '(x /= 0) and (1 /x < y)' when x has the value 0.

ADA is full of features like *and then* and *or else* that you should know of because you are liable to meet them in programs written by others, and you may find them useful in your programs. Some of these features you can find in the syntax diagrams

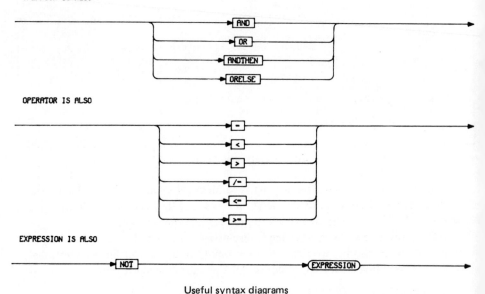

Useful syntax diagrams

We can rewrite our day counting program so that it uses the two new kinds of *loop*

```
with        Text_IO
procedure DayCount is
     use Text_IO;
        type MONTH is (january,february,march,april,may,june,july
                     ,august,september,october,november,december);
     m   :MONTH; y:INTEGER; d:INTEGER range 1..31;i:INTEGER;
     month_length:array(MONTH) of INTEGER;
begin
     month_length := (31,28,31,30,31,30,31,31,30,31,30,31);
     d:=4;m:=december;y:=1937;Get(i);
     --  my birthday can be replaced by your birthday
     while i>0
     loop for n in 1..i -- we do not need to declare n
          loop
               if    d < month_length(m)
               then  d:= d+1;          -- d set to next day
               elsif m < december
               then  d:=1;
                     m:= MONTH'succ(m);-- m set to next month
               else  d:=1;m:=january;
                     y:=y+1;           -- y set to next year
               end   if;
               --   adjust for leap year
               if    y mod 4 /=0 then month_length(february):= 28;
               elsif y mod 100=0 then month_length(february):= 28;
               else                    month_length(february):= 29;
               end   if;
          end loop;
          Put(d);Put(MONTH'pos(m));Put(y);NewLine;Get(i);
     --   'pos converts the value of m to an integer
     end   loop;
end        DayCount;
```

Revised ADA program: Day Count

The most extreme way of limiting the values that may be assigned to a variable is to make it a constant. If we make a declaration like

tax-procent: *constant* INTEGER := 97

then the computer will protest if the program tries to assign to the variable 'tax_ procent'. There lies a moral behind our choice of this variable name. Once the politicians in Denmark were unable to change the tax laws in the way they wanted because a crucial number occurred at several thousand places in the tax programs. If the programmers had introduced this number as a constant and used the constant name instead of the number, the politicians would not have been thwarted.

Exercise

Write a program that finds the day of the week of any given date. Your program should contain the line

type WEEKDAY *is* (sunday, monday, tuesday, wednesday, thursday, friday, saturday);

Exercise

In section 4.1 you met the environment Date Manipulator and its two procedures Create and Interval. Write a program for these two procedures and use Interval to give a short version of our program DayCount.

138

5.4 Correct programs

How can we be confident that a program does in fact solve the problem it is intended to solve? Until now we have assumed that a program is acceptable, when it gives the expected results for well chosen test data. As there is no point in writing a program to solve a problem when every solution is known, we want more than this. In this section we explain how we can prove that a small program is correct, if we assume that the computer does not malfunction and its primitive algorithm Translate does not introduce errors. No large program is correct but one should do what one can to prevent a program from producing an incorrect result. Any way of guaranteeing the correctness of a small part of a large program increases our confidence in the correctness of the large program.

The basic idea is to insert assertions into programs and algorithms. An assertion is a statement about the values of variables of the program or algorithm. At any given time such a statement is either true or false; when we insert it at some place in a program or algorithm, we express the intention that the assertion is always true when the program or algorithm is at that place. Let us look at an example. When you read about the justification problem in section 3.4 and met the structure diagram

JUSTIFY IS

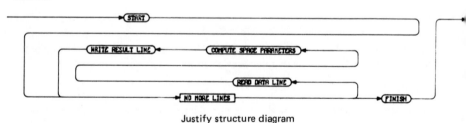

Justify structure diagram

you probably wondered why we had the Start and Finish oblongs. Now we can reveal that they were assertions.

START IS

◄WE CHECK THAT ALL LINES HAVE A SPACE AND THEY ARE NOT TOO LONG►

FINISH IS

◄LEFT AND RIGHT MARGINS ARE NOW ALIGNED►

Two assertions

In the ADA program for the justification problem the Finish assertion appears as a comment and the Start assertion appears as conditions for raising the exceptions No Space and Too Long. Assertions in programs can appear in several ways: as value constraints on variables and parameters, as conditions that raise exceptions.

The assertion idea has become so popular that it illustrates the proverb 'A favourite child has many names'; in other texts assertions are called snapshots, predicates or invariants. With this in mind we can cut the gordian knot at the end of the last chapter, we can use an assertion as the guarantee between an environment and its users. Do you remember the environment

```
package DateManipulator is
        type DATE is limited private;
        function Create(day,month,year:INTEGER) return DATE;
        function Interval(after,before:DATE) return INTEGER;
private type DATE is new INTECER;
end     DateManipulator;
```

Date Manipulator package interface

This environment can satisfy the assertion: Interval (after, before) returns the number of days between the dates after and before so it satisfies the law 'Interval $(a,b) = -$Interval (b,a)'.

When an environment introduces a type, it is a good idea to put laws about operations on the type because such laws can be used to prove the correctness of a program in the same way that the arithmetical law

any_interger $+ 0 =$ any_integer

can be used to prove the equality of $2 \times (3 + 0)$ and 2×3. Remember the problem:

given non-zero numbers $a_1, a_2, \ldots, a_n$ followed by 0
compute the sum $a_1 + a_2 + \ldots + a_n$ and the double sum
$$(a_1) + (a_1 + a_2) + \ldots + (a_1 + a_2 + \ldots + a_n)$$

The natural algorithm for this problem is

ACCUMULATE IS

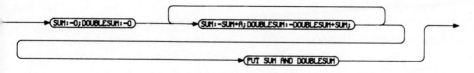

Accumulate structure diagram

and the problem is solved by the program

```
with        Text_IO
procedure Accumulate is
        use Text_IO;
        number,sum,double_sum:INTEGER;
begin
        sum := 0; double_sum := 0;
        loop
        -- sum and double_sum are
        -- correct for numbers read
            Get(number);
            exit when    number=0;
            sum := sum + number;
            double_sum:=double_sum+sum;
        end    loop;
        Put(sum);Put(double_sum);
    end        Accumulate;
```

ADA program: Accumulate

Let us prove that this program is correct because its assertion is always true. Clearly the assertion is true when we enter the loop. Assume that it is true for $m <$ k. When we obey Get(number) for the kth time we have

$\text{sum} = a_1 + a_2 + \ldots + a_{k-1}$

$\text{double\_sum} = (a_1) + (a_1 + a_2) + \ldots + (a_1 + a_2 + \ldots + a_{k-1})$

Obeying 'sum := sum + a_k; double_sum := double_sum + sum;'gives

$\text{sum} = a_1 + a_2 + \ldots + a_k$

$\text{double\_sum} = (a_1) + (a_1 + a_2) + \ldots + (a_1 + a_2 + \ldots + a_k)$

so the assertion is true for $m < k + 1$ and we have proved the correctness of the program by induction.

In this section we have given three arguments for the use of assertions (1) they help in program development, (2) as constraints or conditions they catch errors, (3) so programs can be proved correct. Opinions differ about the relative weights of these arguments, but all authorities agree that assertions are useful.

Exercise

Insert appropriate assertions in the program Sloping Line in section 3.2. Then prove the program correct.

Exercise

Check that the following function gives the result 8 for the data 2, 3 and the result 9 for the data 3, 2. Prove that the function always computes 'c to the power d' when c and d are positive numbers.

```
function FastPower(c,d:INTEGER) return INTEGER is
         a,b,x:INTEGER;
begin
            x := 1;a:=c;b:=d;
            loop
            --     (a to power b)*x is c to power d
                   if     b <= 0
                   then   return x;
                   elsif  b mod 2 = 1
                   then   x := x * a;
                   end    if;
                   a:=a*a;b:=b/2;
            end    loop;
end         FastPower;
```

Fast power function

Exercise

Develop a program for finding the greatest common divisor of two positive numbers from the assertions:

greatest_common_divisor (a, b) = greatest_common_divisor $(a, b\text{-}a)$
greatest_common_divisor (a, b) = greatest_common_divisor (b, a)
greatest_common_divisor $(a, a) = a$

Exercise

In an exercise at the end of the last section you wrote programs for the procedures Create and Interval in the environment. Date_Manipulator. Now prove that your programs satisfy the law Interval(a, b) = -Interval(b, a).

Exercise

Write a program that computes the function

$$f(1) = 1$$
$$f(2k) = k$$
$$f(2k+1) = f(f(3k+2))$$

where k is a positive integer. Prove that for every $x > 1$ such that $f(x)$ is defined we have

$$f(x) \leqslant x/2$$

The first few values of f are

	1	2	3	4	5	6	7
$f(x)$	1	1	1	2	2	3	1

but no one knows if $f(x)$ is defined for all x.

Chapter 6

More on ADA Types

An important part of writing a program to solve a program is to choose an appropriate representation for the program concepts. In this chapter we describe the powerful ADA mechanisms for introducing and using types. We illustrate these mechanisms by completing the programs for the environments introduced in chapter 4.

6.1 Files

Suppose we ask the computer to obey the primitive DO(my_data, my_results, my_code), when the variable my_code contains the translated version of an ADA program with statements Get(a) and Put(b). The program will fetch information from my_data when it obeys the statement Get(a) and it will send information to my_results when it obeys the statement Put(b). Provided my_data is prepared to supply information, and my_results is prepared to receive information, all will be well.

In practice we often want to fetch information from more than one source and we want to send information to more than one destination. Fortunately ADA allows us to declare parameters of type INFILE for information sources and parameters of type OUTFILE for information destinations. We can write procedures like

```
procedure Copy(s:INFILE;d:OUTFILE) is
        c   :CHARACTER;
begin
        loop
                Get(s,c);
                Put(d,c);
        end    loop;
exception when End_Error =>null;
end        Copy;
```

Copy procedure

142

This procedure uses the exception End_Error which is raised when the program tries to fetch non-existing information, so it does in fact copy the values of s to d.

We can also declare variables of type INFILE and OUTFILE, but there is a problem. There must be a connection between the values of these variables when the program is being obeyed and the values of other variables that are external to the program. In our description of the primitives TRANSLATE, EDIT and DO you met external variables, but you have not yet learnt how these variables are created and destroyed. Suppose we have a program that declares a variable internal_file_name of type INFILE or OUTFILE. Before we can use Put or Get on this variable the program must obey one of

— Create (internal_file_name, external_file_name);
  if the variable external_file_name does not exist
— Open (internal_file_name, external_file_name);
  if the variable external_file_name already exists.

Before the program finishes it must obey one of

— Close (internal_file_name);
  if we want to keep the external variable associated with internal_file_name.
— Delete (external_file_name);
  if we want to destroy the external variable associated with external_file_name.

Consider the program

```
with        Text_IO
procedure Copy is
      use Text_IO;
      s  :INFILE ;
      d  :OUTFILE;
      c  :CHARACTER;
begin
          Open(s,"source");        -- "source" exists
          Create(d,"destination"); -- "destination" does not
          loop
              Get(s,c);
              Put(d,c);
          end   loop;
          Delete("source");        -- "source" forgotten
          Close (d);               -- "destination" remembered
exception when End_Error => null;
end       Copy;
```

ADA program: Copy

This program copies the value of the external variable 'source' to the external variable 'destination', if it is obeyed when 'source' exists but 'destination' does not. As you now know enough for practical problem solving, let us say no more on the complicated subject of files in ADA.

Exercise

Write three ADA programs: the first program should make a file of the even numbers less than 1000, the second program should make a file of the multiples of 7 less than 1000, the third program should combine these two files into one.

144

6.2 Strings, sequences and arrays

In many problems we want to use sequences of characters as values, so ADA provides the type STRING for this purpose. The four character sequence 'IRIT' is a value of the type STRING in just the same way that the negative number '−1576' is a value of the type INTEGER. The alert reader will be aware of the fact that ADA must have some convention for representing sequences of characters, because blank characters cannot be seen and sequences can be empty. Instead of explaining this convention in details, we shall illustrate it be examples:

"IRIT"	—— 4 non-blank characters
"WILEY INTERNATIONAL"	—— 19 characters of which 1 blank
" "	—— 1 blank character
""	—— 0 characters
""""	—— 1 double quote character.

The last example illustrates the ADA rule that double quote characters must be repeated if they occur in the string being represented.

The only ADA primitive for combining strings is concatenation, &, placing the strings side by side. The precise definition of concatenation can be reconstructed by the mathematically minded from the fact that the following program

```
with       Text_IO
procedure LoveStory is
        use Text_IO;
        l  :constant STRING := " loves ";

        procedure Question(subject,object:STRING) is
        begin
                Put(subject & l & object & "?");
        end       Question;

begin  -- LoveStory
                Question("John","Mary");
                Question("Mary","John");
                Question("Ursula","");NewLine;
end       LoveStory;
```

ADA program: Love Story

produces the output

John loves Mary? Mary loves John? Ursula loves?

In this program we have introduced a constant and two parameters of type STRING, and we have used the string version of Put.

To consolidate your understanding of string operations and type coercions we give the body of the environment Syntax:

```
with         Text_IO
package body Syntax  is

        -- "type RULE is new STRING;"should replace
        -- "--RULE is specified later" in earlier
        -- "package Syntax is"

        function Oblong(name:LINE) return RULE is
        begin
                return RULE("%"&STRING(name)&"%");
        end     Oblong;

        function Rectangle(name:LINE)return RULE is
        begin
                return RULE("#"&STRING(name)&"#");
        end     Rectangle;

        function "&" (left,right:RULE) return RULE is
        begin
                return RULE(STRING(left)&STRING(right));
        end     "&";

        function "/" (left,right:RULE) return RULE is
        begin
                return RULE(STRING(left)&"~"&STRING(right));
        end     "/";

        function Option(r:RULE) return RULE is
        begin
                return RULE("["&STRING(r)&"]");
        end     Option;

        function Repeat(left,right:RULE) return RULE is
        begin
                return RULE(STRING(left) &
                            "{"&STRING(right)&"}");
        end     Repeat;

        -- the next two subprograms use GetLine and
        -- PutLine from the environment Text_IO

        function GetRule (name:LINE) return RULE is
        begin
                if    STRING(name)=Text_IO.GetLine()
                then  return RULE(Text_IO.Get_Line());
                end   if;
        end     GetRule;

        procedure PutRule(name:LINE;r:RULE) is
        begin
                Text_IO.Put_Line(STRING(name));
                Text_IO.Put_Line(STRING(r));
        end     PutRule;

end          Syntax;
```

Syntax package body

An inelegant feature of ADA is its requirement that private types like RULE are revealed in the environment specification instead of the environment body. The reason for this inelegance is that the computer must be told how much memory space a variable requires, when the variable is declared. You can see how the computer gets this information about string variables in the program

```
with       LineManipulator
procedure  Promiscuous is
      use  LineManipulator;
           boy,girl:STRING(1..80);
begin
           boy  := STRING(DataLine());
           Put(boy);Put(" is interested in");NewLine();
           loop
                girl:= STRING(DataLine());
                exit when boy = girl;
                Put(girl);NewLine;
           end  loop;
end        Promisdcuous;
```

ADA program: Promiscuous

When we give the type STRING (1 . . 80) to the variables boy and girl, we implicitly get new character variables like boy (3) and new string variables like girl (2 . . 4). Implicit variables are used in the procedures

```
with       Text_IO
procedure  Palindrome is
      use  Text_IO;
           boy,girl:STRING(1..5);
begin
           for   n in 1..5
           loop
                Get(boy(n));
                girl(6-n)   := boy(n);
           end  loop;
           Put(boy);Put("reversed is"); Put(girl);
           if    boy=girl
           then Put(" : a palindrome");
        -- like the introduction of the first man
        -- to the first woman "Madam,I'm Adam"
           end  if;NewLine;
end        Palindrome;
```

```
with       LineManipulator
procedure  Marriage is
      use  LineManipulator;
           boy,girl:STRING(1..80);
begin
           boy  := STRING(DataLine());
           girl := STRING(DataLine());
           Put(girl);NewLine;
           Put("changes her name to :");NewLine;
           girl(41..80):=boy(41..80);
           Put(girl);NewLine;
end        Marriage;
```

ADA programs: Palindrome and Marriage

Implicit variables like boy (41 .. 80) are known as "sliced" variables; they were used in the procedure Marriage to avoid changing a girl's surname character by character. As you would probably like to see a larger example of the use of implicit variables, we give the body of the environment Line Manipulator.

```
with         Text_IO
package body LineManipulator is
         -- in the interface we introduced
         -- limit,spaces:INTEGER
         -- type LINE is new STRING(1..80)
         subtype POSITION is INTEGER range 0..80;
         index:POSITION;l:LINE;c:CHARACTER;

         procedure Insert(old-line:LINE) is
         begin
                   l:=old-line;index:=0;
         end       Insert;

         function NextCharacter return CHARACTER is
         begin
                   if   index = 80 then return LF;end if;
                   index:= index+1;      return l(index);
         end       NextCharacter;

         function LineToCharacter return CHARACTER is
         begin
                   loop
                        c := NextCharacter();
                        case c is
                             when 'a'..'z'|LF => return c;
                             when others      => null;
                        end  case;
                   end  loop;
         end       LineToCharacter;

         function  Keyboard return INFILE is
         -- gives the internal file name for Your display keyboard
         -- Because your ADA system probably does not allow you to
         -- submit programs with our Do primitive,you may have to
         -- redefine this function
         begin
                   return Standard_Input();
         end       Keyboard;

         function DataLine return LINE is
         begin
                   Set_Input(Keyboard());
                   limit:=0;spaces:=0;index:=0;
                   l:=(1..80=>LF);
                   loop
                        Get(c);
                        exit when c=LF;
                        if   limit < 80
                        then limit:= limit +1;
                             l(limit) := c ;
                             if   c=' '
                             then spaces:=spaces+1;
                             end  if;
                        end  if;
                   end  loop;
                   index:= 0;
                   Set_Input(Current_Input());
                   return l;
         end       NextLine;
end      LineManipulator;
```

Line Manipulator package body

148

In many problems we want to work with sequences and use operations like concatenation and slicing. In ADA we can introduce types by declarations like

type VECTOR is array (INTEGER range <>) of INTEGER

and we can write procedures like

```
procedure Largest(v:in VECTOR;position,max:out INTEGER) is
begin
        position := v'FIRST;max:=v(position);
        for i in v'FIRST + 1..v'LAST
        loop
                if v(i)>max
                then   position := i;max:=v(i);
                end if;
        end   loop;
end       Largest;
```

Typical array procedure

Before you can write programs that work with sequences, you need to know how to associate sequence types with variables and how to represent a value of a sequence type. This you can learn from the somewhat artifical program

```
with        Text_IO
procedure VectorGeneration is
        use Text_IO;
        type  VECTOR is array(INTEGER range<>) of INTEGER;
        v   :  VECTOR(0..20);
begin
        v(0):=1;v(1):=1;
        for i in 2..20
        loop
                v(i):=v(i-1)+v(i-2);
                Put(v(i));
        end   loop;
end        VectorGeneration;
```

ADA program: Vector Generation

Sometimes it is more natural to work with limited sequences instead of unlimited. You can see how this is done in the body of the environment Text Manipulator

```
with        LineManipulator
package body TextManipulator is
        -- TEXT is array(1..80) of LINE should replace
        -- "-TEXT is specified later" in earlier
        -- "package TextManipulator is"
        -- user has exception NoMoreInput
        lindex,pindex:POSITION;
        l:LINE;c:CHARACTER;
        function Empty return TEXT is
        begin
                return(1..80=>(1..80=>LF));
        end   Empty;
        function Number     return INTEGER is
                -- unknown to user
            m :INTEGER;
        begin
                m := 0;
                loop
                    c := NextCharacter();
                    case c          is
                        when '0'   => m:=m*10;
                        when '1'   => m:=m*10+1;
                        when '2'   => m:=m*10+2;
                        when '3'   => m:=m*10+3;
                        when '4'   => m:=m*10+4;
                        when '5'   => m:=m*10+5;
                        when '6'   => m:=m*10+6;
```

```
                           when '7'   => m:=m*10+7;
                           when '8'   => m:=m*10+8;
                           when '9'   => m:=m*10+9;
                           when others=> return m ;
                   end  case;
             end  Loop;
     end       Number;

     procedure Correct(t:in out TEXT) is
           m   :INTEGER;
     begin
             l := DataLine();
             lindex:=Number();
             if    c = ':'
             then  t(lindex):= DataLine();
             else  m:= Number();
                   t(lindex..m):=(lindex..m =>(1..80=>LF));
             end if;
     exception when others => raise NoMoreInput;
     end       Correct;

     procedure PutText(t:TEXT) is
     begin
             lindex := 1;
             Loop
                   if   t(lindex)(1) =/ LF
                   then PutLine(t(lindex)  );
                   end  if;
                   exit when lindex = 100;
                   lindex := lindex + 1;
             end  Loop;
     end       PutText;

     function  Screen return OUTFILE is
               -- comment like Keyboard in LineManipulator
     begin
             return Standard_Output();
     end       Screen;

     procedure Renumber(t: in out TEXT) is
           u   :TEXT;m:INTEGER;
     begin
             lindex := 1;m:=2;u:=Empty();
             Set_Output(Screen());
             Loop
                   if   t(lindex)(1) =/ LF
                   then Put(m);PutLine(t(lindex));
                        u(m):=t(lindex);m:=m+2;
                        NewLine;
                   end  if;
                   exit when lindex = 80 or m=80;
                   lindex := lindex + 1;
             end  Loop;
             Set_Output(Current_Output());
             t := u;
     end       Renumber;

     function  GetText return TEXT is
           t   :TEXT;
     begin
             t:=Empty();
             lindex:= 2;
             Loop
                   GetLine(t(lindex));
                   lindex:=lindex + 2;
             end  Loop;
     exception when others => return t;
     end       GetText;

end       TextManipulator;
```

Text Manipulator package body

So far we have only described how one can work with one dimensional sequences in ADA. It is just as easy to work with multidimensional sequences; you only need to know the syntax diagrams

DECLARATION IS ALSO

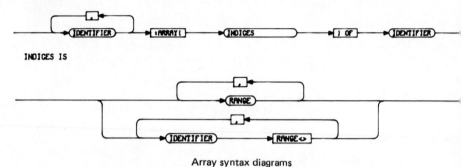

INDICES IS

Array syntax diagrams

and the fact that ADA does not allow concatenation and slicing for dimensions greater than one.

Exercise

Devise a program that finds the rhyming pattern of a poem. Two lines of the poem get the same letter in the rhyming pattern if their endings sound the same. Your program should find the rhyming pattern ABAB in the Ogden Nash poem

I test my bath before I sit,
And I'm always moved to wonderment
That what chills the finger not a bit
Is so frigid upon the fundament.

but it may have difficulty find the limeric pattern AABBA in

```
The left sock of a certain professor



Was greater, the right one was lesser.



He said 'Though irrational,



It must be quite fashionable



I've a similar pair in the dresser!
```

Rhyming Test data

6.3 Records

In the last section you saw how variables of the same type can be combined into one variable of *array* type; in this section you will see how variables of different types can be combined into a single variable in ADA. If we make declarations like

```
type PERSON is record
                 first name, surname: STRING (1 .. 10);
                 birthday: DATE;
                 sex: (MALE, FEMALE);
           end record;
you, me: PERSON;
```

we not only introduce two variables of type PERSON, but we also introduce eight implicit variables: you. first name, you. surname, you. birthday, you. sex, me. first name, me. surname, me. birthday, me. sex. These implicit variables can be used as if they had been declared separately; we can freely make assignments like

me. first name	:= "ANON";
me. surname	:= me. first name;
me. birthday	:= Create (30, 2, 1980); — using Date Manipulator
me. sex	.= MALE;

and we can change the values of several implicit variables at the same time by an assignment like 'you := me;'. You will become convinced that record types are useful when you study the body of the environment Private Data Base:

```
package body PrivateDataBase
-- uses package TableManipulator specified in
-- the exercise at the end of chapter 7.
-- "type RELATION is record
--                       names    : LINE;
--                       contents:TABLE;
--                 end       record; "
-- should replace "type RELATION is given later"
-- in earlier "package PrivateDataBase is"
        function "*"(first,second:RELATION) return RELATION is
        begin
                return( names   => first.names,
                        contents=> Mix
(first.contents,second.contents,product,first.names=second.names));
        end      "*";
        function "+"(first,second:RELATION) return RELATION is
        begin
                return( names   => first.names,
                        contents=> Mix
(first.contents,second.contents,sum,first.names=second.names));
        end      "+";
        function "-"(first,second:RELATION) return RELATION is
        begin
                return( names   => first.names,
                        contents=> Mix
(first.contents,second.contents,differ,first.names=second.names));
        end      "-";
        function Join  (first,second:RELATION) return RELATION is
        begin
                return( names   => MatchOnFirstName(first.names,second.names),
                        contents=> Mix
                                (first.contents,second.contents,join,TRUE));
        end      Join;
        function Project(r:RELATION;title:LINE) return RELATION is
        begin
                return( names   => Permute(r.names,title),
                        contents=> Modify (r.contents,title,project));
        end      Project;
        function Create(title:LINE) return RELATION is
        begin
                return( names   => title,
                        contents=> Convert(DataLine()));
        end      Create;
        function Same  (r:RELATION) return RELATION is
        begin
                return( names   => r.names,
                        contents=> Modify(r.contents,DataLine(),same));
        end      Same;
        function Before(r:RELATION) return RELATION is
        begin
                return( names   => r.names,
                        contents=> Modify(r.contents,DataLine(),before))
        end      Before;
        function After (r:RELATION) return RELATION is
        begin
                return( names   => r.names,
                        contents=>Modify(r.contents,DataLine(),after));
        end      After;
        function GetRelation return RELATION is
        r:RELATION;
```

```
            begin
                    r.title := LINE(GetLine());-- GetLine is a Text_IO function
                                            -- that gives a string
                    GetTable(r.contents,Occurrences(".",r.title)+1);
                    return r;
            end     GetRelation;
            procedure PutRelation(r:RELATION) is
            begin
                    PutLine(STRING(r.names));-- PutLine is a Text_IO procedure
                                          -- that requires a string
                    PutTable(r.contents);
            end     PutRelation;
    end     PrivateDataBase;
```

Private Data Base package body

As you write more and more ADA programs using record types, you will feel certain needs. In the next section we will feel the need for a way of defining closely related record types. We shall satisfy this need by declaring an ADA record type with a discriminant, we shall write

type ELEMENT (size: INTEGER range 1 .. 10) *is*
    record
          value: array (1 .. size) of STRING (1 .. 10);
          above, below: TABLE;
    end record;

If we associate the type ELEMENT (2) with a variable e we would get five implicit variables:

e. value, e. value (1), e. value (2), e. above, e. below

If we had associated the type ELEMENT (20) with another variable, it would have had 23 implicit variables; the discriminant 'size' determines a particular record type in a family. As another example of a record type with a discriminant consider

```
type GENDER is (MALE,FEMALE);

type PERSON (sex:GENDER:=FEMALE)
            is  record
                            firstname,surname:STRING(1..10);
                            birthday :DATE;
                            case sex  is
                                    when    MALE    => weight:INTEGER;
                                    when    FEMALE => pregnant:BOOLEAN;
                            end case;
                end   record;
you:PERSON;         -- you.sex is initially female
me: PERSON(MALE); --  me.sex is always male
```

Typical discriminated records

This introduces several implicit variable including:

you. sex, you. weight, you. pregnant, me. sex, me. weight, me. pregnant

but there are restrictions on the way these variables can be used:

  — you. weight is only meaningful when you. sex = MALE
  — you. pregnant is only meaningful when you. sex = FEMALE

— me. pregnant is never meaningful
— me. sex always has the value MALE
— you. sex can only be changed by an assignment to the variable 'you'.

The computer will always check that these requirements are met; the translator will raise the exception CONSTRAINT ERROR if it meets the assignment 'me := you' when you. sex = FEMALE.

There is a very convenient way of assigning values to ADA variables of record type—one can assign an aggregate like

```
(sex=>FEMALE,firstname=>"Ada",surname=>"Lovelace",
 birthday=>Create(29,2,1800) ,pregnant=>FALSE)
```

Typical record aggregates

to the variable you. This assignment gives a value to the implicit variable you. sex, this value can be changed by assignments like 'you := me' but assignments like 'you. sex := MALE' are not allowed. Aggregates can also be used with ADA variables of array type. If we have the declarations

type WEEKDAY is (SUNDAY, MONDAY, TUESDAY, WEDNESDAY,
           THURSDAY, FRIDAY, SATURDAY)
A: array (WEEKDAY) of INTEGER;

then we can assign any of the aggregates

```
(0,8,8,8,8,8,0)
(SUNDAY=>0,   MONDAY..FRIDAY=>8,  SATURDAY=>0)
(SUNDAY!SATURDAY=>0,MONDAY..FRIDAY=>8)
```

Typical array aggregates

to the variable A instead of writing seven separate assignment statements.

Exercise

In the last section we gave the body of the environment Date Manipulator. Write a new version that assumes

type DATE is record
                  day: INTEGER *range* 1 . . 31;
                  month: INTEGER *range* 1 . . 12;
                  year: INTEGER;
           end record

All programs using Date Manipulator should work with both versions of the environment body.

6.4 Access types

More and more governments are introducing unique identifications of their citizens. As citizens are born and die these identification systems have to provide far more identifications than tne actual number of citizens alive on any particular day. This situation often arises when writing a program to solve a problem—we want to identify and work with a varying number of objects of type T. Fortunately, ADA has an 'access type' for any type, and the values of *access* T are identifications of values of type T. Thus the declarations

type PERSON_IDENTIFIER *is access* PERSON;
         her, him: PERSON_IDENTIFIER; me: PERSON;

give a type, whose values are different from values of type PERSON. You do not need to know what the values of PERSON_IDENTIFIER really are, only that they can be used to identify persons. New values of PERSON_IDENTIFIER are created when we obey assignments like

her := new PERSON (sex = > FEMALE, . . .);
him := new PERSON (me);

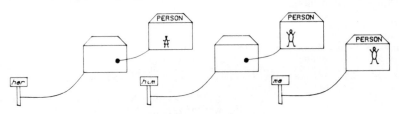

Creation of access type values

We can copy values of PERSON_IDENTIFIER from one variable to another, and we can note that a variable is not being used to identify a person:

her := him ; him := null;

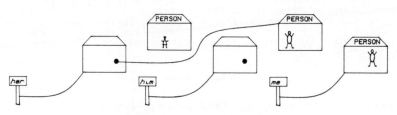

Assignment to access type variables

As you probably expect, declaring a variable of access type also gives several implicit variables; our declarations of him and her give the implicit variables

- him. first name, him. surname, her. first name, her. surname
        of type STRING (1 . . 10)
- him. birthday, her. birthday of type DATE
- him. sex   , her. sex   of type GENDER
- him. *all*   , her. *all*   of type PERSON

which we can use in assignments like

156

her. all := him. all

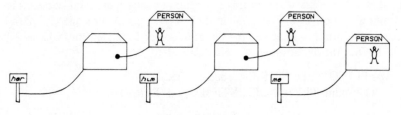

Copying of accessed values

Note the subtle difference between 'her. all := him. all' and 'her := him'; the first assignment does not change the value of 'her', the second does.

Now we can explain how we can work with a dynamically varying number of objects. Suppose we are trying to solve a problem about groups of people. If we introduce the environment

```
with      DateManipulator
package PersonManipulator is
          type PERSON;
          type PERSON-IDENTIFIER is access PERSON;
          type PERSON-NAME is array(INTEGER range<>) of CHARACTER;
          type PERSON  is record
                              firstname,surname:PERSON-NAME(1..30);
                              birthday:DateManipulator.DATE;
                              sex:(MALE,FEMALE);
                              spouse:PERSON-IDENTIFIER;
                     end     record;
          -- we will use  the following functions and procedures
          -- in later programs but you can define them as you wish
          function Somebody                 return PERSON-IDENTIFIER;
          function Neighbour   (pi:PERSON-IDENTIFIER)
                                             return PERSON-IDENTIFIER;
          procedure PutFriends  (pi:PERSON-IDENTIFIER);
          function Younger(a,b:PERSON-IDENTIFIER) return  BOOLEAN;
          function Identified(pi:PERSON-IDENTIFIER;p:PERSON)
                                             return  BOOLEAN;
          function Christen   return  PERSON-NAME;
end       PersonManipulator;
```

Package interface

we can easily program the build up of multiperson families like

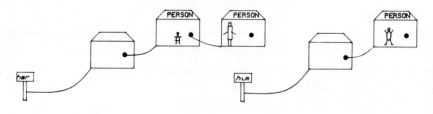

Copying of access values

Obeying the assignment 'Him. spouse := Her' in this situation would give

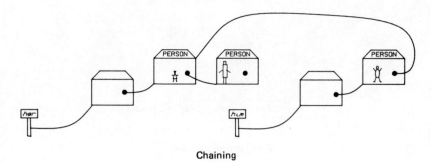

Chaining

and we could refer to the square fellow by either of the two implicit variables:
Her. spouse, Him. spouse. spouse.

The normal reaction of a programmer to dynamic data types, such as ADA's
access types, is a refusal to believe that they can be useful in practice. To avoid this
reaction we shall give the body of the environment Drawing.

```
package body Drawing is
        --      type NODE;
        --      type FIGURE is access NODE;
        --      type NODE is record
        --                              l,d,a:INTEGER;
        --                                  n:FIGURE;
        --                  end         record;
        --      should replace "FIGURE is specified later"
        --      in earlier specification of "package Drawing"

        function Line(length,direction:INTEGER) return FIGURE is
        begin
                if length<0 then return null;end if;
                return new NODE(l=>length,d=>direction,a=>0,n=>null);
        end     Line;
        function Arc (length,direction,angle:INTEGER) return FIGURE is
        begin
                if length<0 then return null;end if;
                return new NODE(l=>length,d=>direction,a=>angle,n=>null);
        end     Arc;
        function Catenate(left,right:FIGURE) return FIGURE is
        begin
                if   left=null then return right;end if;
                return new NODE(l=>left.l,d=>left.d,a=>left.a,
                                n=>Catenate(left.n,right));
        end     Catenate;
        function Scale(f:FIGURE;procent:INTEGER) return FIGURE is
        begin
                if   f=null then return null;end if;
                return new NODE(l=>f.l*procent/100,d=>f.d,a=>f.a,
                                n=>Scale(f.n,procent));
        end     Scale;
        function Shift(f:FIGURE;length,direction:INTEGER) return FIGURE is
        begin
                if f=null  then return null;end if;
                return new NODE(l=>-length,d=>direction,a=>0,n=>f);
                -- line with negative length is invisible!
        end     Shift;
        function Rotate(f:FIGURE;angle:INTEGER) return FIGURE is
        begin
                if   f=null
                then return new NODE(l=>0,d=>0,a=>0,n=>null);
                end if;
                return new NODE(l=>f.l,d=>f.d+angle,a=>f.a,
                                n=>Rotate(f.n,angle));
        end     Rotate;
        function GetFigure return FIGURE is
                ll,dd,aa:INTEGER;f:FIGURE;
```

```
begin
        loop
                Get(ll);Get(dd);Get(aa);
                f:= new NODE(l=>ll,d=>dd,a=>aa,n=>f);
        end     loop;
exception when End_Error=>return f;
end     GetFigure;
procedure PutFigure(f:FIGURE) is
begin
        if  f=null then return;end if;
        PutFigure(f.n);Put(f.l);Put(f.d);Put(f.a);
end     PutFigure;
end     Drawing;
```

Drawing package body

6.5 Types in the computer

When a computer obeys a program, it keeps the values of the program variables in its store. You can think of the computer store as a large number of cells, each of which has an address. When you declare a variable of type INTEGER, CHARACTER or BOOLEAN, the variable will be assigned one cell in the store

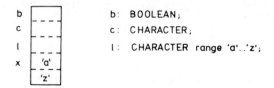

```
b                       b:  BOOLEAN;
c                       c:  CHARACTER;
l                       l:  CHARACTER range 'a'..'z';
x       'a'
        'z'
```

Simple variables in the computer

In this picture you see that variable names become cell addresses and type information disappears. The primitive TRANSLATE, which converts an ADA program into a form the computer can obey, uses the type information to check for errors in the program.

```
i:=2;           --  Load            (2)
                --  Store In  Cell(i)
l:='c';         --  Load            (c)
                --  Check Range     (x)
                --  Store In  Cell(l)
l:= c;          --  Load From Cell(c)
                --  Check Range     (x)
                --  Store In  Cell(l)
c:='l';         --  Load            (l)
                --  Store In  Cell(c)
```

Translation of simple variable statement

Notice the 'run-time check' for the subtype variable, and the subtle differences between Load (1) and Load From Cell (1).

When we use the primitives TRANSLATE, DO, and EDIT, we also use variables that are external to all ADA programs. Obeying these primitives changes the values of the external variables:

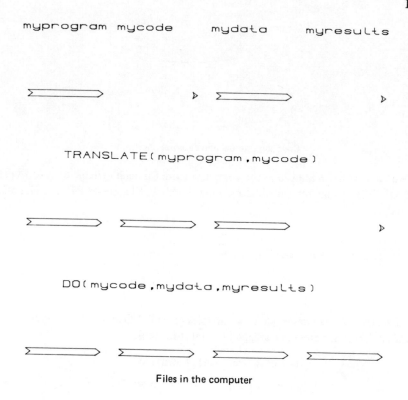

Files in the computer

An ADA program can fetch information from an external value by obeying a Get statement, and it can change the value of an external variable by obeying a Put statement.

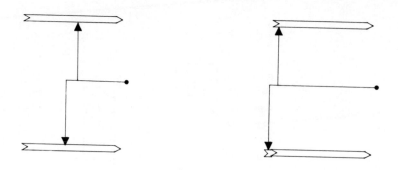

Put and Get making changes

In section 6.1 we described how program variables of type INFILE and OUT-FILE can be connected to external variables

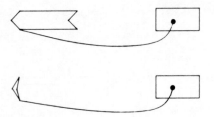

Coupling internal and external file names

We see that there is a cell in the computer store for each variable of type INFILE and OUTFILE, and we can describe how TRANSLATE treats Put and Get statements:

```
Get(inf,i);      --  Load From File(inf)
                 --  Store In   Cell(i)
Put(outf,o);     --  Load From Cell(o)
                 --  Store In   File(outf)
```

Translation of put and get statements

In section 6.2 we described how variables of STRING and ARRAY types can be declared. These variables are assigned several store cells:

```
boy,girl : array (1..3) of MYSTERY

page     : array (1..2,1..3) of CHARACTER
```

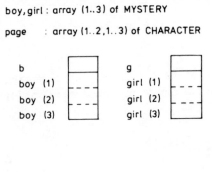

Arrays in the computer

The primitive TRANSLATE has to insert many run time checks, when it is given a program with string and array variables.

```
boy(1):=girl(2);    --  Load              (2)
                    --  Check Range       (x)
                    --  Store In  Cell(g)
                    --  Load              (1)
                    --  Check Range       (x)
                    --  Store In  Cell(b)
                    --  Load From Cell(girl,g)
                    --  Store In  Cell(boy,b)

i:=page(2,3);       --  Load              (3)
                    --  Check Range       (z)
                    --  Store In  Cell(p)
                    --  Load              (2)
                    --  Check Range       (y)
                    --  Multiply          (3)
                    --  Add    To  Cell(p)
                    --  Load From Cell(page,p)
                    --  Store In  Cell(i)
```

Translation of array statements

You should be aware of these run-time checks because they make programs slow and expensive; sometimes you may want to transform a program to a more efficient, if less readable, form.

```
if      boy(i)          -- LoadFromCell(i)      c:= boy(i);    -- LoadFromCell(i)
                        -- CheckRange   (x)                    -- CheckRange   (x)
                        -- StoreInCell  (b)                    -- StoreInCell  (b)
   mod girl(2)          -- Load              (2)               -- LoadFromCell(boy,b)
                        -- CheckRange   (x)                    -- StoreInCell (c)
                        -- StoreInCell  (g)  if c mod girl(2)-- Load            (2)
                        -- LoadFromCell(boy,b)                 -- CheckRange   (x)
    /= 0                -- ModWithCell(girl,g)                 -- LoadFromCell(c)
                        -- ZeroTest (....)                     -- ModWithCell(girl,g)
then  return boy(i)     -- LoadFromCell(i)       /=0           -- ZeroTest(.....)
                        -- CheckRange   (x)   then return c    -- LoadFromCell(c)
                        -- StoreInCell  (b)                    -- Return
                        -- LoadFromCell(boy,b)
                        -- Return
```

Avoiding run time checks

In section 6.3 we described how variables of RECORD type can be declared. These variables are also assigned several store cells:

```
boy, girl : record
          first,second,third: MYSTERY
          end record;
```

Records in the computer

Record variables usually give more efficient programs than array variables because they do not require address calculations and run-time checks.

```
if       boy.first          -- Load From Cell( boy,1)
      mod girl.second       -- Mod  With Cell(girl,2)
 then return boy.first      -- Load From Cell( boy,1)
                            -- Return
```

Translation of record statements

In section 6.4 we described how variables of access type can be declared. An access variable is assigned one store cell, and the value of such a variable is a cell address.

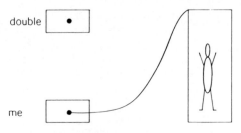

Access types in the computer

The cell addresses, that are values of access variables, are never of any interest. When a program obeys a statement like

double := new PERSON (me)

it is given a number of cells where it can copy the value of the person me, and it places the addresses of the first of these cells as the value of the access variable her

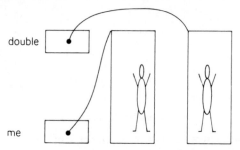

Creation of access type values

Now that you know how store cells are assigned to variables, we can give the reasons for some of the ADA features that you may have found puzzling. It is often convenient to use parameters of unlimited type

function Maximum (v: VECTOR) return INTEGER

and we can do this because the computer does not need to know how many store cells to assign to a parameter until the associated procedure or function is called.

In practice it is often convenient to translate parts of a program separately; when we make a small change in a program, we want to retranslate as little of the program as possible. Instead of translating an environment every time we translate a program that uses the environment, we would like to save time and money by translating the environment only once. In ADA we can translate a program P that begins 'with E', whenever the interface part of E has been translated. Because the program P can use environment variables and types, the primitive TRANSLATE must know

— the address of the cells assigned to environment variables

— how many cells to assign to a variable of environment type.

This is the reason why private types are revealed in the interface of an environment (and not in the body where they intuitively belong).

At any given time the programmer will have a library of translated interfaces, bodies and programs. She can add to this library by obeying

TRANSLATE (P, c)

when P is the body of an environment whose interface is in the library, *or* P does not require an interface that is not yet in the library. If this translation is successful, the programmer can now obey

DO (her_data, her_results, c)

when P is a program *and* the bodies of all environments required by P have been translated.

Exercise

Choose a program and an environment in this chapter. Assign store cells to each variable and translate some of the program statements.

6.6 Tasks

In ADA a task is a special kind of variable which you cannot assign to. To emphasize this, we list the interfaces of all the tasks in this book (except that in section 4.5 for the task Public Data Base):

```
          -- tasks in section 3.5
   task  A  is
                           entry Suck;
   end   A;

   task  B  is
                           entry Blow;
   end   B;

   task  Plotter is
                           entry Plot;
                           entry Release;
   end   Plotter;

          --  tasks in this section
   task type SEMAPHORE is
                           entry P;
                           entry V;
   end           SEMAPHORE;

   task type PLAYER;

          --  tasks in next chapter
   task Stacktask is
                           entry Push(e:in  ITEM);
                           entry Pop;
                           entry Top  (e:out ITEM);
   end   StackTask;
```

```
task BufferTask is
                        entry Add     (i:in  ITEM);
                        entry Delete(i:out ITEM);
end  BufferTask;

task type  T is
                        entry Read  (i:out ITEM);
                        entry Write (i:in  ITEM);
end  T;

task Externals is
                        entry Make      (s:STRING);
                        entry Destroy  (s:STRING);
                        entry OneIn     (s:STRING;e:in  ELEMENT);
                        entry OneOut    (s:STRING;e:out ELEMENT);
                        entry ManyIn    (s:STRING;v:in  VECTOR );
                        entry ManyOut   (s:STRING;v:out VECTOR );
                        entry MakeP     (p:PROFILE);
                        entry DestroyP(p:PROFILE);
                        entry Advance  (p:PROFILE);
end Externals;

task RunAdministrator is
                        entry Corrupt     (a,c:STRING;  runs:out  INTEGER);
                        entry TwoRunMerge (a,b,c:STRING;aend,bend:BOOLEAN);
                        entry ManyRunMerge(p:PROFILE;c:STRING);
                        entry Kill        (       a,c:STRING);
end RunAdministrator;

task TwoFile is
                entry TwoSplit      (x,y,z  :STRING);
                entry SimpleMerge    (x,y,z  :STRING;     runs:out INTEGER);
                entry BalancedMerge  (x,y,u,v:STRING;     runs:out INTEGER);
                entry FibonacciMerge(x,y,z  :STRING;     runs:out INTEGER);
end  TwoFile;

task ManyFile is
                entry ManySplit(p:PROFILE;from:STRING);
                entry MergeSimple(p:PROFILE;to:STRING;   runs:out INTEGER);
                entry MergeBalanced(from,to:PROFILE;     runs:out INTEGER);
                entry MergeFibonacci(p:PROFILE;to:STRING;runs:out INTEGER);
end  ManyFile;
```

ADA task interfaces

When we described the ADA task concept in section 3.5, we introduced network diagrams with coloured oblongs. The declaration of a task variable corresponds to making a WHITE oblong. A WHITE oblong becomes GREEN when a program reaches the *begin* after the task declaration. There are two ways a GREEN oblong can become WHITE

—— no more activity because the task has reached the end of its body
—— the task has obeyed a *terminate* statement.

Sometimes one wants to wait for a task to become WHITE before doing something else, sometimes one does not. The ADA convention is to wait for a task variable at the *end* of the construction that declared the task variable.

```
task type TableAdministrator is
     -- in next chapter's package TableManipulator
procedure P is
          a:TableAdministrator;
          procedure Q is
                    b:TableAdministrator;
          begin            -- task b activated
                    null;
               --   statements for Q
          end       Q; -- wait for b to stop
begin                -- task a activated
          Q;
end       P;              -- wait for a to stop
```

Starting and Stopping of Tasks

Now we turn to the more exciting side of ADA tasks: the role of entries in task bodies. In the body of the task we can have *accept* statements:

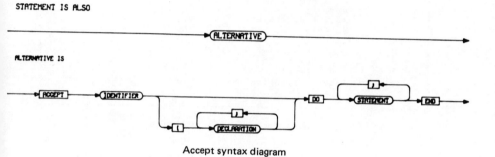

Accept syntax diagram

When the active task meets such a statement, it waits for a rendezvous on the entry (GREEN to YELLOW); when the rendezvous happens the statements between *do* and *end* are obeyed (YELLOW to GREEN).

```
with      Text-IO
procedure ResultDrivenPipeline is
          n:INTEGER;data,result:CHARACTER;
          task A is entry Suck;end A;
          task body A is
          begin
               loop
                  select accept Suck do
                                UseData;
                         end;-- rendezvous over
                  or delay  60.0
--see ADA manual for why 60.0 not 60 and package Calendar
                         Put("Minute Wasted");
                  or terminate;
--chosen when program wants to stop
                  end select;
                  exit when n=0;
                  n:= n-1;
               end  loop;
          end A;
begin  -- task A activated
          Get(n);
          loop
                Get(data);Suck;Put(result);
                exit when n=0;
          end  loop;
end       ResultDrivenPipeline;
```

ADA program: Data Driven Pipeline

We often want a task to wait for possible rendezvous on several entries at the same time, so we use a *select* statement

STATEMENT IS ALSO

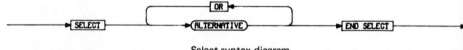

Select syntax diagram

When the active task meets such a statement it waits for a rendezvous on any of the entries in the *accept* statements

```
task body RunAdministrator
begin
     loop
          select  accept Corrupt(a,c:STRING;runs:out INTEGER) do ... end;
             or   accept TwoRunMerge(a,b,c:STRING;
                                     aend,bend:out BOOLEAN) do ... end;
             or   accept ManyRunMerge(p:PROFILE;c:STRING)    do ... end;
             or   accept Kill(a,c:STRING) do ... end; -- rendezvous over
                  exit;
          end  select;
     end  loop;
end       RunAdministrator;
```

Run Administrator task body

ADA provides several ways of getting a task to do something useful instead of waiting a long time for some rendezvous.

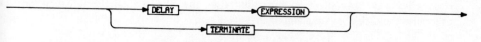

Terminate and Delay syntax diagrams

The meaning of *delay* and *terminate* alternative should be clear from the example

```
task body Externals is
          stored,revealed:INTEGER:=0;
begin
          loop
               select terminate;
                   or delay(60.0);Put("stored and released since last report")
                                    Put(stored);Put(released);NewLine;
                                    stored:=0;released:=0;
                   or accept Make(s:STRING)                      do...end;
                   or accept Destroy(s:STRING)                   do...end;
                   or accept OneIn(s:STRING;e:in ELEMENT)        do...end;
                                    stored:=stored +1;
                   or accept OneOut(s:STRING;e:out ELEMENT) do...end;
                                    released:=released +1;
                   or accept ManyIn(s:STRING;v:in VECTOR)        do...end;
                                    stored:=stored +500;
                   or accept ManyOut(s:STRING;v:out VECTOR)      do...end;
                                    released:=released +500;
                   or accept MakeP   (p:PROFILE)               do...end;
                   or accept DestroyP(p:PROFILE)               do...end;
                   or accept Advance (p:PROFILE)               do...end;
               end  select;
          end  loop;
end       Externals;
```

Externals task body

There is one more feature of ADA select statements, you should know about: conditional alternatives.

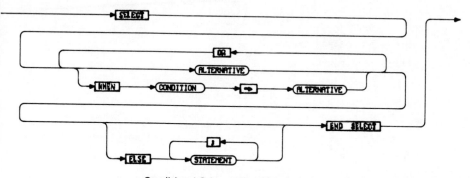

Conditional Select syntax diagram

When the active task meets a *select* statement with conditional alternatives, it evaluates all the conditions and decides if there is an open alternative. An open alternative is a select alternative without a condition or one with a condition that evaluates to TRUE.

If there is an open alternative, the task waits for a rendezvous; if there is no open alternative and no *else,* the task raises the SELECT_ERROR exception; otherwise the task obeys the statements after *else.*

```
task body Plotter is
        busy:BOOLEAN;
begin
        busy:=FALSE;
        loop
            select when not Busy=>
                    accept Plot  do
                            busy:=TRUE;
                    end;
                or when not busy=>
                    terminate;
                or accept Release do
                            busy:=FALSE;
                    end;
            end    select;
        end    loop;
end    Plotter;
```

Plotter task body

Have you noticed that ADA allows several tasks to share global variables? This is powerful but dangerous, and you should know how to protect yourself from the dangers. Suppose we have

```
procedure Soccer is
        game_over:BOOLEAN;
        task type SEMAPHORE is
                    entry P;
                    entry V;
        end     SEMAPHORE;
        task type PLAYER;
        ball:SEMAPHORE;
        us,them:array(1..11) of PLAYER;

task body SEMAPHORE is
begin
        loop
                select terminate
                    or accept P do end;
                end    select;
                accept V do end;
        end    loop;
end     SEMAPHORE;

task body PLAYER is
begin
        loop
                exit when game_over;
                TakeUpPosition;
                ball.P ; -- critical region
                Dribble; -- two players cannot have
                ball.V ; -- the ball at the same time
        end    loop;
end     PLAYER;
-- if you want to keep the players moving
-- consult the ADA manual about variants of select

begin    -- 23 tasks activated
        game_over:=FALSE;
        delay(90*60.0);
        game_over:=TRUE;
end     Soccer;
```

Semaphore discipline

If we have a critical region in our task body, a group of statements that refers to shred variables, we can protect it by putting semaphore. P before the region and semaphore. V after. You should check that this way of using a semaphore ensures that one never has two tasks in a critical region at the same time.

Exercise

Program a data-driven pipeline. In this section we gave a result-driven pipeline.

Exercise

Write the task body of Public_Data_Base. In section 6.4 we gave the package body of Private_Data_Base.

Chapter 7

Sorting, Searching and other Examples of ADA Generics

In the same way that roses and violets are instances of the same genera Flowers, ADA allows different subalgorithms as instances of the same generic subalgorithm. In this chapter you will see many generic subalgorithms and you will learn how the activity of ADA procedures and tasks can be represented in a computer. In the last section you will find an extensive discussion of the important topics of sorting and searching.

7.1 Generic procedures and functions

In many problems we want two variables of the same type to interchange their values. Instead of doing this by a Swap procedure for each type, we can write the generic procedure

```
generic     type ITEM is private;
procedure   Swap(a,b:in out ITEM);

procedure   Swap(a,b:in out ITEM) is
       c  : ITEM;
begin
            c:=a;a:=b;b:=c;
end         Swap;
```

Typical generic procedure

The primitive Translate will use this to create an instance procedure whenever it means lines like

procedure Swap Integer is new Swap (INTEGER);
procedure Swap Character is new Swap (CHARACTER);
procedure Swap Vector is new Swap (VECTOR);

instead of writing several procedures with very similar bodies, we write one generic procedure and several instantiations. The idea of replacing several procedures by one generic procedure is very convenient because ADA allows procedures and functions as generic parameters

```
generic     type ITEM is <>;
            with function Earlier(a,b:ITEM) return BOOLEAN;
function    Max(a,b:ITEM) return ITEM;

function    Max(a,b:ITEM) return ITEM is
begin
            if    Earlier(a,b)
            then  return b;
            else  return a;
            end   if;
end         Max;
```

Function as generic parameter

The primitive Translate will use this to create instance functions whenever it means lines like

function Max Integer is new Max (INTEGER, "<");
function Max Character is new Max (CHARACTER, "<");
function Max Vector is new Max (VECTOR, Before);

the effect is the same as if we had written three very similar functions

```
function MaxInteger(a,b:INTEGER) return INTEGER is
begin
        if    a < b
        then  return b;
        else  return a;
        end   if;
end        MaxInteger;

function MaxCharacter(a,b:CHARACTER) return CHARACTER is
begin
        if    a < b
        then  return b;
        else  return a;
        end   if;
end        MaxCharacter;

function MaxVector(a,b:VECTOR) return VECTOR is
begin
        if    Before(a,b)
        then  return b;
        else  return a;
        end   if;
end        MaxVector;
```

Instances of a generic function

Note that the definition of the function Max Vector presupposes that the type VECTOR and the function Before are defined already.

Exercise

Fill in the body of the generic function

generic (type ELEMENT
        function "x" (a, b: ELEMENT) return ELEMENT);
function Power (e: ELEMENT; n: INTEGER) return ELEMENT);
        -- returns e multiplied by itself n times

Check that function Mult is new Power (INTEGER, "+") is the same as ordinary integer multiplication when n is positive. Modify your generic function so that it obeys the law

Power (a, n) X Power (a, −n) = unit_element

You can assume an inverse function '/' as a new generic parameter.

7.2 Generic packages

Many packages become much more useful, if they are given generic parameters. Hitherto we have used ADA's 'with' construction to give meaning to the 'global names' in a package, procedure or function, but we could also have specified these 'global names' as generic parameters. The interface of a generic package looks like

```
generic      type ITEM is private;
package      Stack       is
             type LIST is private;
             procedure Create(l: out LIST);
             procedure Push(l:in out LIST;i:in ITEM);
             procedure Pop( l:in out LIST);
             function Top(l:LIST) return ITEM;
             Underflow:EXCEPTION;
private      type NODE;
             type LIST is access NODE;
             type NODE is record
                                       head:ITEM;
                                       tail:LIST;
                             end record;
end          Stack;
```

Stack generic package interface

The primitive Translate will use this to give a package interface when it meets lines like

package Character Stack is new Stack (CHARACTER);
package Word Stack is new Stack (Character Stack, LIST);

The effect is the same as if we had written the two package interfaces

```
package CharacterStack is
        type LIST is private;
        procedure Create(l: out LIST);
        procedure Push(l:in out LIST;i:CHARACTER);
        procedure Pop (l:in out LIST);
        function  Top (l:LIST) return CHARACTER;
        Underflow:EXCEPTION;
private type NODE;
        type LIST is access NODE;
        type NODE is record
                                head:CHARACTER;
                                tail:LIST;
                    end record;
end     CharacterStack;

package WordStack is
        type LIST is private;
        procedure Create(l: out LIST);
        procedure Push(l:in out LIST;i:CharacterStack.LIST);
        procedure Pop (l:in out LIST);
        function  Top (l:LIST)  return CharacterStack.LIST;
        Underflow:EXCEPTION;
private type NODE;
        type LIST is access NODE;
        type NODE is record
                                head:CharacterStack.LIST;
                                tail:LIST;
                    end record;
end     WordStack;
```

Instances of a generic package

As you might expect the values of the type Word Stack. LIST are lists of lists of characters, and we can represent them as trees of depth 2

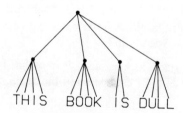

Typical stack value

All instances of our generic package STACK acquire a body when we provide

```
package body Stack is
        procedure Create(l:out LIST) is
        begin               l:= null;
        end     Create;

        procedure Push(l:in out LIST;i:ITEM) is
        begin
                l:=new LIST(head=>i,tail=>l);
        end     Push;

        procedure Pop (l:in out LIST) is
        begin
                if    l = null
                then raise Underflow;
                else l := l.tail;
                end  if;
        end     Pop;

        function Top (l:LIST) return ITEM is
        begin
                if    l = null
                then raise Underflow;
                else return l.head;
                end  if;
        end     Top;
end     Stack;
```

Stack package vbody

Notice that

— Word Stack. Create   returns an empty Word Stack. LIST
— Word Stack. Push(e, 1)   adds the alue of e to the list l
— Word Stack. Top (1)   returns the first element of l
— Word Stack. Pop(e)   deletes the first element of l
— the exception Word Stack. underflow is raised when Top or Pop is applied to an empty Word Stack. LIST.

Two variants of our stack package are often useful. In the first variant new elements are added at the end of a list instead of the beginning

```
generic     type ITEM is private;
package     Queue is
            type LIST is private;
            procedure Create(l:    out LIST);
            procedure Add    (l:in out LIST;i:in  ITEM);
            procedure Delete(l:in out LIST;i:out ITEM);
            Underflow:EXCEPTION;
private     type NODE;
            type LIST is access NODE;
            type NODE is record
                                        head:ITEM;
                                        tail:LIST;
                        end record;
end         Queue

package body Queue is
        procedure Create(l:out LIST) is
        begin
                l:=null;
        end     Create;

        procedure Add(l:in out LIST;i:in ITEM) is
        begin
                l:=new LIST(head=>i,tail=>l);
        end     Add;

        procedure Delete(l:in out LIST;i:out ITEM) is
        begin
                if    l = null
                then  raise Underflow;
                elsif l.tail=null
                then  i:=l.head;l:=null;
                else  Delete(l.tail,i);
                end   if;
        end     Delete;
end     Queue;
```

Queue generic package

In the second variant we limit the size of lists by a new kind of generic parameter

```
generic    size:INTEGER;
           type ITEM is private;
package    Block is
           type VECTOR is private;
           procedure Create(v:out VECTOR;initial:ITEM);
           procedure Change(v:in out VECTOR;n:INTEGER;i:ITEM);
           function  Occurs(v:VECTOR;i:ITEM) return INTEGER;
           OutOfRange:EXCEPTION;
private    type VECTOR is array(1..size) of ITEM;
end        Block;

package body Block is
           procedure  Create(v:out VECTOR;initial:ITEM) is
           begin
                      v:=(1..size=> initial);
           end        Create;

           procedure  Change(v:in out VECTOR;n:INTEGER;i:ITEM) is
           begin
                      if    n in 1..size
                      then  v(n) := i;
                      else  raise OutOfRange;
                      end   if;
           end        Change;

           function   Occurs(v:VECTOR;i:ITEM) return INTEGER is
           begin
                      n:= size;
                      loop
                             if    t(n) = i
                             then  return n;
                             elsif n = 1
                             then  return 0;
                             else  n := n-1;
                             end   if;
                      end loop;
           end         Occurs;
end        Block;
```

Block generic package

The way one instantiates packages with the new kind of generic parameters is as you would expect: the lines

package Character Block is new Block (20, CHARACTER);
package Word Block is new Block (100, Character Block. List);

give packages for manipulating up to 100 words, each consisting of 20 characters or less.

Stacks and queues can be used to solve many problems (if you do not feel like meeting a rather difficult example, you can skip the rest of this section). We can use stacks to explain what happens in a computer when programs call and return from procedures and functions. Suppose we have a type ACTIVATION_RECORD whose values are 'values of variables local to a procedure or function'. If we define

package Simulate is new Stack (ACTIVATION_RECORD);

we get

— — Simulate. Create   corresponds to starting a program
— — Simulate. Push   corresponds to calling a procedure or function

—— Simulate. Pop   corresponds to returning from a procedure or function

—— Simulate. Underflow   corresponds to finishing a program.

If we have a program P that calls a procedure Q and then later a function R, then we might have the following sequence of values of Simulate. LIST

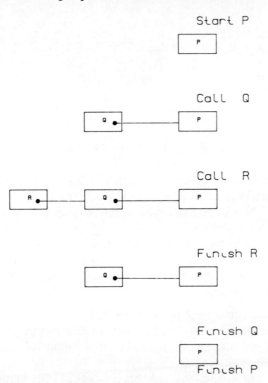

Executing a sequential program

Suppose our program P runs concurrently with a number of tasks. Each task can call and return from procedures and functions, so we will have a sequence of values of Simulate. LIST for each task

Main program

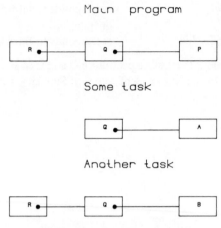

Some task

Another task

Executing a concurrent program

It is natural to think of values of Simulate. LIST as 'threads of control' and summarise our explanation: there is a thread of control for the main program and each active task, these grow and shrink as the program and tasks call and return from procedures and functions.

You may be wondering whether threads of control can clarify what happens when two tasks make a rendezvous with one another. In the same way that we assigned colours to the tasks in a network diagram in section 3.5, we can assign colours to threads of control

— a white thread corresponds to an inactive task;
— a green thread corresponds to a task that can do something;
— a yellow thread corresponds to a task that is waiting for some other task to call one of the task's entries;
— a red thread corresponds to a task that is calling an entry in some other task.

The basic algorithm for implementing tasks on a computer is given by the structure diagram

CONCURRENT ADA IS

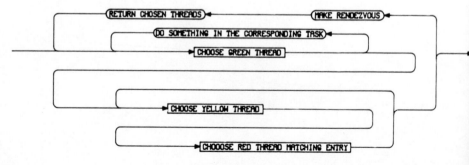

Concurrent execution structure diagram

This diagram suggests the use of lists of threads

— green list for the list of green threads;
— yellow list for the list of yellow threads;
— red list E for the list of red threads calling entry E.

Because ADA insists that entry calls are queued, the obvious way of introducing these lists of threads is

package Parallel is new Queue (Simulation. LIST);

Now we can use the package functions to refine our structure diagram

CHOOSE GREEN THREAD IS

CHOOSE YELLOW THREAD IS

Choosing a Green or Yellow task

The function Parallel.Add will be useful in refining 'return chosen threads', but none of the package functions will help with 'make rendezvous' or 'do something in the corresponding task'. Perhaps the reader would like to see the refinement of 'choose red thread matching entry':

CHOOSE RED THREAD MATCHING ENTRY IS

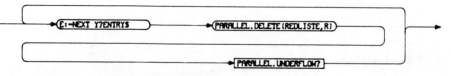

Choosing a Red task

In the hope of clarifying the ADA task concept we have sketched one possible ADA implementation, but there are others. ADA does not require queueing of green and yellow threads, and it allows truly parallel implementations for those fortunate enough to have multi-computer systems.

Exercise

Design a generic package for trees of arbitrary depth and width. In this section you have seen trees of depth 2 and arbitrary width.

Exercise

Design a generic package for trees of arbitrary depth and width 2. Do not use your solution to the previous exercise, but make something more efficient.

Exercise

The programs in this book have used the Put and Get operations of the ADA package Text10. This package begins with an instantiaton InputOutput(CHARACTER) of a generic package that gives Write and Read operations. Your exercise is to define the Text10 operations for Putting and Getting integers from character files. From what you have seen so far you might imagine that the InputOutput interface looks like

```
generic type ELEMENT   is limited private;
package INPUT-OUTPUT   is
        type   IN-FILE is limited private;
        type OUT-FILE is limited private;
        procedure Create(file:in out   IN-FILE;name:in STRING);
        procedure Create(file:in out OUT-FILE;name:in STRING);
        procedure Open   (file:in out   IN-FILE;name:in STRING);
        procedure Open   (file:in out OUT-FILE;name:in STRING);
        procedure Close (file:in out   IN-FILE);
        procedure Close (file:in out OUT-FILE);
        procedure Delete(name:in STRING);
        procedure Read   (file:in   IN-FILE);
        procedure Write (file:in OUT-FILE);
private -- omitted
end      INPUT-OUTPUT;
```

Simple INPUT OUTPUT generic package interface

Try to approximate the actual InputOutput interface by allowing for

— a type INOUTFILE for files on which one can both read and write;
— an operation IsOpen for finding out if a file is open or not;
— an operation Name for finding the external name of a file;
— an operation Size for finding the number of elements in a file.

7.3 Generic tasks

The choice between a sequential solution of a problem and a parallel solution is unaffected by the need for a generic solution. As the following variant of Stack shows, the parallel solution of a problem may be simpler than the sequential solution because concurrency avoids tedious exceptions.

```
generic    type ITEM is private;
package    StackManipulator is
           task StackTask    is
               entry Push(e:in   ITEM);
               entry Pop ;
               entry Top (e:out ITEM);
           end   StackTask;
end        StackManipulator;

package body StackManipulator is
           type NODE;
           type LIST is access NODE;
           type NODE is record
                               head:ITEM;
                               tail:LIST;
                    end   record;
           task body StackTask is
               l:LIST;
           begin
                l:=null;
                loop
                    select   accept Push(e:in ITEM) do
                             l:= new NODE(head=>e,tail=>l);
                             end;
                    or       when l/=null  =>
                             accept Pop do
                             l:=l.tail;
                             end;
                    or       when l/=null  =>
                             accept Top(e:out ITEM) do
                                  e := l.head;
                             end;
                    or       terminate;
                    end   select;
                end loop;
           end   StackTask;
end        StackManipulator;
```

Generic Stack Task

You should know about the following variant of Queue because it is often useful in the task solution of a problem.

```
generic     size:INTEGER;
            type ITEM is private;
package     BufferManipulator is
            task BufferTask    is
                entry Add(i:in ITEM);
                entry Delete(i:out ITEM);
            end  BufferTask;
end         BufferManipulator;

package body BufferManipulator is
            task body BufferTask    is
                buffer:array(1..size) of ITEM;
                count,first,last:INTEGER;
            begin
                count:=0;first:=1;last:=1;
                loop
                    select when count<size =>
                            accept Add(i:in ITEM) do
                                    buffer(first):= i;
                            end;
                            count:=count+1;
                            first:=first mod size +1;
                        or  when count>0      =>
                            accept Delete(i:out ITEM) do
                                    i:=buffer(last);
                            end;
                            count:=count-1;
                            last := last mod size +1;
                        or  terminate;
                    end select;
                end loop;
            end  BufferTask;
end         BufferManipulator;
```

Generic Buffer Task

Task solutions of problems have their own difficulties, because many tasks can share the same resource. You have already seen the use of semaphores to protect critical regions in co-operating tasks. Sometimes an instance of the generic task

```
generic  type ITEM is private;
package  ProtectedVariable is
         task type T is
               entry Read(i:out ITEM);
               entry Write(i:in ITEM);
         end  T;
end      ProtectedVariable;

package  body ProtectedVariable is
         task body T is
               v:ITEM;
         begin
                     select
                             accept Write(i:in ITEM) do
                                    v:=i;
                             end;
                     or      terminate;
                     end     select;
               loop
                     select
                             accept Write(i:in ITEM) do
                                    v:=i;
                             end;
                     or      accept Read(i:out ITEM) do
                                    i:=v;
                             end;
                     or      terminate;
                     end     select;
               end   loop;
         end   T;
end      ProtectedVariable;
```

Generic Task for protecting shared variables

gives a more convenient way of protecting critical regions, because other tasks can
only access the protected variables one at a time.

Exercise

Modify the generic task Protected Variable so that it allows several readers to look at
the value of the variable concurrently, but it still gives writers exclusive access. Does
your solution force writers to wait a long time when they are many eager readers?
If so, modify your solution so that no new reader is admitted when a writer is
waiting.

Exercise

Design a generic package Post Office with a task that accepts and delivers letters.
You can assume that it is the responsibility of the Post Office users to put letters in
public post boxes and collect letters from their private homes, so the interface of
Post Office is:

```
generic type LETTER is private;
        with function Address(l:LETTER) return STRING;
package PostOffice is
        type MESSAGE;
        type BUNDLE is access MESSAGE;
        type MESSAGE is record
                                l:LETTER;
                                b:BUNDLE;
                        end     record;
        task PostalService is
            entry NewHouse(where:STRING);
            entry PutInPostBox(post:BUNDLE);
            entry FetchPost(where:STRING);
        end   PostalService;
end       PostOffice;
```

Post Office generic package interface

In the operating systems of many computers there is a program similar to Post Office.

7.4 Sorting and searching

At any given time a large proportion of the world's computers are searching in their memories or sorting information for future searchs—one estimate is 20%. Many sorting and searching algorithms have been devised, each with their virtues and vices; in this section we give several algorithms, because so many problems require sorting and searching and there is no algorithm that is always best. We shall start with searching methods that are *internal* because they do not manipulate *external* files. We continue with internal sorting methods, external searching methods and external sorting methods. We finish with *mixed* methods because external methods must be used with the large files that arise in practice, but an external method can be improved dramatically by combining it with an internal method.

Let us begin with an algorithm for finding the position of the least element in a structure

LEAST IS

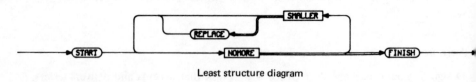

Least structure diagram

A common way of programming this algorithm is to choose an array type for structures

```
generic type ITEM is (<>);
        --   because we have <> not private
        --   ITEM is scalar type and
        --   we can use <,>,<=,>=
        type VECTOR is array(INTEGER range<>) of ITEM;
function  Least(a:VECTOR) return ITEM;

function  Least(a:VECTOR) return ITEM is
        m:ITEM;n:INTEGER;
begin
        n:=a'FIRST;m:=a(n);
        loop
            if    n =a'LAST
            then  return m;
            elsif a(n+1)<m
            then  m:=a(n+1);
            end   if;
            n:=n+1;
        end loop;
end       Least;
```

Least generic function

We get another refinement of our algorithm if we choose an access type for structures

```
generic type NODE is private;
        type LIST is access NODE;
        with function"<"(a,b:LIST) return BOOLEAN;
        with function Next(c:LIST) return LIST;
function  LeastList(l:LIST)      return LIST;

function  LeastList(l:LIST)      return LIST is
    m   :LIST;
begin
        if    l=null
        then  return l;
        else  m:= LeastList(Next(l));
            if    l<m
            then  return l;
            else  return m;
            end   if;
        end   if;
end       LeastList;
```

Least List generic function

You can see instances of our two generic procedures in

```
with      PersonManipulator, --from chapter 6
          Generics   -- everything in this chapter
procedure SmallPeople is
      use PersonManipulator;
          function LeastCharacter is  Generics.
                        new Least(CHARACTER,PERSON_NAME);
          function Youngest is Generics.
                        new LeastList(PERSON,PERSON_IDENTIFIER,
                                          Younger,Neighbour);
begin
          Put("Least Character is");
          Put(LeastCharacter(Christen()));NewLine;
          PutFriends(Youngest(Somebody()));
end       SmallPeople;
```

Instances of Least and Least List

Notice that Least Character returns a CHARACTER, while Youngest returns a PERSON IDENTIFIER.

Searching algorithms differ from algorithms for finding least elements because they may not find the element they are looking for, but they can stop looking if they do find it. We get our first internal searching algorithm by modifying the structure diagram for Least

SEARCH UNSORTED

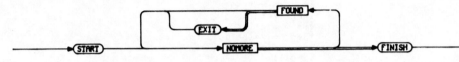

Search Unsorted structure diagram

If we choose an array type for structures, this gives the generic function

```
generic    type ITEM is (<>);
           type VECTOR is array(INTEGER range<>) of ITEM;
function   SearchUnsorted(v:VECTOR;i:ITEM) return INTEGER;

function   SearchUnsorted(v:VECTOR;i:ITEM) return INTEGER is
       n :INTEGER;
begin
           n:= v'FIRST;
           loop
               if    v(n) = i
               then  return n;
               elsif n=v'LAST
               then  return 0;
               else  n:=n+1;
               end   if;
           end  loop;
end        SearchUnsorted;
```

Search Unsorted generic function

If we choose an access type for structures, we get a similar generic function

```
generic    type NODE is private;
           type LIST is access NODE;
           with function Next(l:LIST) return LIST;
           with function Match(l:LIST;n:NODE) return BOOLEAN;
function   UnsortedSearch(l:LIST;n:NODE) return LIST;

function   UnsortedSearch(l:LIST;n:NODE) return LIST is
begin
           loop
               if    l=null
               then  return null;
               elsif Match(l,n)
               then  return l;
               else  UnsortedSearch(Next(l),n);
               end   if;
           end  loop;
end        UnsortedSearch;
```

Unsorted Search generic function

You can see instances of these two generic functions in

```
with        PersonManipulator, --from chapter 6
            Generics    -- everything in this chapter
procedure SearchPeople is
     use PersonManipulator;
            function SearchC is new Generics.
                     SearchUnsorted(CHARACTER,PERSON_NAME);
            function SearchP is new Generics.
                     UnsortedSearch(PERSON,PERSON_IDENTIFIER
                                    ,Neighbour,Identified);
            c:CHARACTER;
begin
            Put("Choose Character : ");Get(c);
            Put("Character occurs in position : ");
            Put(SearchC(Christen(),c));NewLine;
            PutFriends(SearchP(Somebody(),Somebody().all));
     end    SearchPeople;
```

Instances of Search Unsorted and Unsorted Search

It is much easier to find information in structures that are sorted—imagine trying to find a word in a dictionary or a telephone number in a directory if they were not sorted. The obvious algorithm for searching a sorted structure is given by

SEARCH SORTED IS

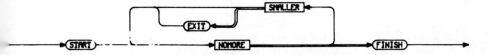

Search Sorted structure diagram

If we choose an array type for structures, this gives the generic function

```
generic    type ITEM is (<>);
           type VECTOR is array(INTEGER range<>) of ITEM;
function   SearchSorted(v:VECTOR;i:ITEM) return INTEGER;

function   SearchSorted(v:VECTOR;i:ITEM) return INTEGER is
        n  :INTEGER;
begin
           n:= v'FIRST;
           loop
                   if     v(n) = i
                   then   return n;
                   elsif  (i<v(n)) or(n=v'LAST)
                   then   return 0;
                   else   n:=n+1;
                   end    if;
           end    loop;
    end    SearchSorted;
```

Search Sorted generic function

If we choose an access type for structures, we get a similar generic function

```
generic    type NODE is private;
           type LIST is access NODE;
           with function Next(l:LIST) return LIST;
           with function Match (l:LIST;n:NODE) return BOOLEAN;
           with function Before(l:LIST;n:NODE) return BOOLEAN;
function   SortedSearch(l:LIST;n:NODE) return LIST;

function   SortedSearch(l:LIST;n:NODE) return LIST is
begin
           loop
                   if    (l=null) or Before(l,n)
                   then  return null;
                   elsif Match(l,n)
                   then  return l;
                   else  SortedSearch(Next(l),n);
                   end   if;
           end   loop;
end        SortedSearch;
```

Sorted Search generic function

We have a much faster way of searching in array structures

```
generic    type ITEM is (<>);
           type VECTOR is array(INTEGER range<>) of ITEM;
function   BinarySearch(v:VECTOR;i:ITEM) return INTEGER;

function   BinarySearch(v:VECTOR;i:ITEM) return INTEGER is
           l,m,r :INTEGER;
begin
           l:= v'FIRST;r:=v'LAST;
           loop
                   m   := (l+r)/2;
                   if    i < v(m)
                   then  if    m=l
                         then  return 0;
                         else  r := m-1;
                         end   if;
                   elsif i > v(m)
                   then  if    m=r
                         then  return 0;
                         else  l := m+1;
                         end   if;
                   else  return m;
                   end   if;
           end   loop;
end        BinarySearch;
```

Binary Search generic function

and we have a much faster way of searching in access structures

```
generic      type NODE is private;
             type TREE is access NODE;
             with function Left    (t:TREE) return TREE;
             with function Right   (t:TREE) return TREE;
             with function After   (t:TREE;n:NODE) return BOOLEAN;
             with function Before  (t:TREE;n:NODE) return BOOLEAN;
function     TreeSearch (t:TREE;n:NODE)      return TREE;

function     TreeSearch (t:TREE;n:NODE)      return TREE is
begin
             if    t=null
             then  return null;
             elsif After(t,n)
             then  return TreeSearch(Left(t),n);
             elsif Before(t,n)
             then  return TreeSearch(Right(t),n);
             else  return t;
             end   if;
end          TreeSearch;
```

Tree Search generic function

A little thought gives the average number of comparisons made by the various internal search methods for a structure with a thousand elements

1000   for Search Unsorted and Unsorted Search
 500   for Search Sorted and Sorted Search
about 10   for Binary Search and Tree Search

With this in mind we follow the designers of a recent IBM computer, aimed at the data base market, and we fix the representation of a data base relation

type TABLE is access ELEMENT;
—— ELEMENT as defined in the last chapter

This choice is wise because users of a data base spend most of their time searching relations, and the tempting alternative of an array representation should be rejected because relations grow and shrink.

Let us begin our discussion of internal sorting methods with sorting by insertion: inserting elements at the appropriate place in a growing structure in the same way that a card player sorts his hand.

INSERT SORT IS

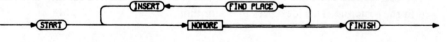

Insert Sort structure diagram

The array structure version of our algorithm is given by

```
generic     type ITEM is (<>);
            type VECTOR is array(INTEGER range<>) of ITEM;
procedure   InsertSort( v:in out VECTOR);

procedure   InsertSort( v:in out VECTOR) is
       l    :INTEGER;
            procedure FindPlace(r:INTEGER) is
            begin
                    if   r = v'FIRST
                    orelse  v(r-1) <= v(l)
                    then v(r):=v(l);
                    else v(r):=v(r-1);
                         FindPlace(r-1);
                    end  if;
            end     FindPlace;
begin
            for     l in v'FIRST+1..v'LAST
            loop -- sorted until position l-1
                    FindPlace(l);
            end     loop;
end         InsertSort;
```

Insert Sort generic procedure

You will understand how this generic procedure works if you see an instance in action

```
with      Generics,PersonManipulator
procedure Sort1 is
        use PersonManipulator;
        procedure Sort is new Generics.
                         InsertSort(CHARACTER,PERSON_NAME);
        n  :PERSON_NAME(1..30);m:INTEGER;
begin
        n := Christen();
        Sort(n);
        for m in 1..30 loop Put(n(m));end loop;NewLine;
end     Sort1;
```

ADA program: Sort 1

Some people find sorting by selection more natural than sorting by insertion. The idea behind sorting by selection is: select the least unsorted element and place it in the structure. This idea gives the algorithm

SELECT SORT IS

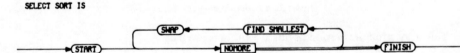

Select Sort structure diagram

and the array structure refinement

```
generic     type ITEM is (<>);

            type VECTOR is array(INTEGER range<>) of ITEM;
procedure SelectSort(v:in out VECTOR);

procedure SelectSort(v:in out VECTOR) is
     l,m:INTEGER;i:ITEM;
            procedure Smallest(r:INTEGER) is
            begin
                      if     l=r
                      then   i:=v(m);v(m):=v(l);v(l):=i;
                      else   if     v(r) < v(m)
                             then   m := r;
                             end    if;
                             Smallest(r-1);
                      end    if;
            end    Smallest;
begin
            for l in v'FIRST..v'LAST
            loop  -- sorted until position l-1
                  m:=v'LAST;
                  Smallest(v'LAST);
            end    loop;
end         SelectSort;
```

Select Sort generic procedure

You will understand how this generic procedure works if you see an instance in action.

```
with      Generics,PersonManipulator
procedure Sort2 is
     use PersonManipulator;
            procedure Sort is new Generics.
                            SelectSort(CHARACTER,PERSON_NAME);
     n   :PERSON_NAME(1..30);m:INTEGER;
begin
            n := Christen();
            Sort(n);
            for m in 1..30 loop Put(n(m));end loop;NewLine;
end         Sort2;
```

ADA program: Sort 2

Our next internal sorting method is popularly known as bubble sorting, because the way it moves elements of a structure until they meet a smaller element reminds one of the way bubbles rise to the top of a water tank.

BUBBLE SORT IS

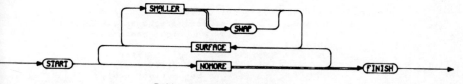

Bubble Sort structure diagram

The array structure refinement of this is

```
generic     type ITEM is (<>);
            type VECTOR is array(INTEGER range<>) of ITEM;
procedure BubbleSort(v:in out VECTOR);

procedure BubbleSort(v:in out VECTOR) is
        l:INTEGER;i:ITEM;
            procedure Bubble(r:INTEGER) is
            begin
                    if      l=r
                    then    return;
                    elsif v(r-1)>v(r)
                    then    i:=v(r-1);v(r-1):=v(r);v(r):=i;
                    end    if;
                    Bubble(r-1);
            end        Bubble;
begin
            for l in v'FIRST..v'LAST
            loop  -- sorted until position l-1
                    Bubble(v'LAST);
            end     loop;
        end        BubbleSort;
```
                    Bubble Sort generic procedure

Again we give an instance in action

```
with      Generics,PersonManipulator
procedure Sort3 is
            use PersonManipulator;
                procedure Sort is new Generics.
                            BubbleSort(CHARACTER,PERSON_NAME);
            n  :PERSON_NAME(1..30);m:INTEGER;
begin
            n := Christen();
            Sort(n);
            for m in 1..30 loop Put(n(m));end loop;NewLine;
        end        Sort3;
```

                    ADA program: Sort 3

so that you can see how bubble sorts differ from insertion sort and selection sort.

The last of our internal sorting methods is called Quick Sort because it usually sorts large arrays much faster than other methods. Our generic procedure Quick Sort assumes we can divide a structure into a non-empty left and right part in such a way that no element in the left part is larger than an element in the right part. This is the role of the procedure Partition in

```
generic     type ITEM is (<>);
            type VECTOR is array(INTEGER range<>) of ITEM;
procedure QuickSort(v:in out VECTOR);

procedure QuickSort(v:in out VECTOR) is
            procedure Sort(left_begin,right_end:INTEGER)is
                        left_end,right_begin:INTEGER;
            begin
                    Partition; -- given later
                    Sort(left_begin,left_end);
                    Sort(right_begin,right_end);
            end     Sort;
begin
            Sort(v'FIRST,v'LAST);
        end        QuickSort;
```
                    Quick Sort generic procedure

If you see an instance of this generic procedure in action

```
with      Generics,PersonManipulator
procedure Sort4 is
      use PersonManipulator;
          procedure Sort is new Generics.
                         QuickSort(CHARACTER,PERSON_NAME);
          n   :PERSON_NAME(1..30);m:INTEGER;
begin
          n := Christen();
          Sort(n);
          for m in 1..30 loop Put(n(m));end loop;NewLine;
end       Sort4;
```

ADA program: Sort 4

you should have no difficulty in programming a suitable Partition procedure

```
procedure Partition is
          --  named unit   assuming variables
          --  left_begin,left_end,right_begin,right_end
              bar,i:ITEM;
              procedure RightMove is
              begin
                        if      v(right_begin) < bar
                        then    right_begin:=right_begin+1;
                                RightMove;
                        end     if;
                  end       RightMove;
                  procedure LeftMove  is
                  begin
                        if      v(left_end) > bar
                        then    left_end := left_end -1;
                        end     if;
                  end       LeftMove;
          begin
              bar:= v((left_begin + right_end)/2);
              left_end:=right_end;right_begin:=left_begin;
              loop
                  RightMove;LeftMove;
                  exit  when left_end < right_begin;
                  i:= v(left_end);
                  v(left_end):= v(right_begin);
                  v(right_begin):= i;
              end   loop;
          end       Partition;
```

Partition procedure

We can use this Partition procedure to give a neat solution of the problem of finding the kth smallest element, in an unsorted structure

```
generic type ITEM is (<>);
         type VECTOR is array(INTEGER range<>) of ITEM;
function Find(v:in out VECTOR;k:INTEGER) return ITEM;

function Find(v:in out VECTOR;k:INTEGER) return ITEM is
         left_begin,left_end,right_begin,right_end:INTEGER;
begin
         left_begin := v'FIRST;right_end := v'LAST;
         loop
               Partition; -- given earlier
               if     k < left_end
               then   right_end := left_end;
               elsif k > right_begin
               then   left_begin:= right_begin;
               else   return v(k);
               end    if;
         end   loop;
end      Find;
```

Find generic function

If we give an instance of this generic function in action

```
with       PersonManipulator, --from chapter 6
           Generics   -- everything in this chapter
procedure LittlePeople is
      use PersonManipulator;
           function P  is new Generics.
                     Find(CHARACTER,PERSON_NAME);
begin
           Put("Next Least Character is");
           Put(P(Christen(),2));NewLine;
end        LittlePeople;
```

ADA program: Little People

you may be tempted to *prove* that Partition and Find always work. The keen reader may also want to prove that all our sorting methods are correct, particularly the non-intuitive Quick Sort.

Before we can describe external searching and sorting methods, we need an environment that provides more powerful file operations than Put and Get. Using the ADA package InputOutput one can write an environment that allows one to put and get several file elements at once and to forget the distinction between INFILE and OUTFILE. We shall assume that the interface of this environment is

```
generic type ELEMENT is private;
        stop,pause:ELEMENT;
        with function Wait(e:ELEMENT) return BOOLEAN;
        with function Key (e:ELEMENT) return INTEGER;
package SortSearch is -- in package Generics
        type VECTOR  is array(1..500) of ELEMENT;
        type PROFILE(length:INTEGER)
            is record
                        name:array(1..length) of STRING(1..10);

                        el  :array(1..length) of ELEMENT;
                end     record;
        task Externals is
            entry Make    (s:STRING);
            entry Destroy(s:STRING);
            entry OneIn  (s:STRING;e:in  ELEMENT);
            entry OneOut (s:STRING;e:out ELEMENT);
            entry ManyIn (s:STRING;v:in  VECTOR );
            entry ManyOut(s:STRING;v:out VECTOR );
            entry MakeP   (p:PROFILE);
            entry DestroyP(p:PROFILE);
            entry Advance (p:PROFILE);
        end    Externals;
-- following package components are defined later
        function FileSearch(f:STRING;n:INTEGER) return INTEGER;
        function FastFileSearch(f:STRING;n:INTEGER) return INTEGER;
        procedure  MergeTwo (a,b,c:STRING);
        procedure  MergeMany(p:PROFILE;c:STRING);
        procedure  SimpleSort        (from,to:STRING);
        procedure  BalancedSort       (from,to:STRING);
        procedure  FibonacciSort      (from,to:STRING);
        -- task      RunAdministrator
        -- task      TwoFile
        -- task      ManyFile
end     SortSearch;
```

Sort Search generic package

and exploit the fact that operations OneOut and ManyOut return stop elements
instead of raising the End—Error exception.

The obvious way to search an external file is closely related to the internal
method of sequential search

```
function    FileSearch(f:STRING;n:INTEGER) return  ELEMENT is
            e:ELEMENT;
begin
            loop
                    OneOut(f,e);
                    if    Key(e)=n or e=stop
                    then  return e;
                    end   if;
            end loop;
end         FileSearch;
```

File Search

Another way of searching a sorted external file is

```
function    FastFileSearch(f:STRING;n:INTEGER) return  ELEMENT is
            v:VECTOR;
begin
            loop
                 ManyOut(f,v);
                 for i in 1..size
                 loop
                          if    Key(e(i))=n or e(i)=stop
                          then  return' e(i);
                          end   if;
                  end  loop;
            end  loop;
end         FastFileSearch;
```

Fast File Search

It is probable that Fast File Search is much faster than File Search because the operations Many Out and One Out take almost the same time.

We now turn to the concept of merging that underlies all external sorting methods. The result of merging the three files

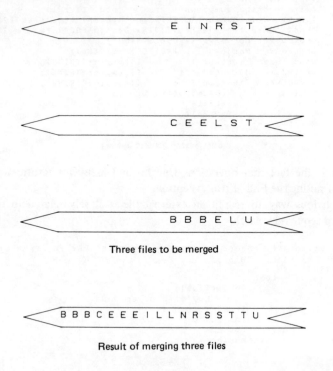

Three files to be merged

is the file

Result of merging three files

We define the result of merging files $f_1, f_2, \ldots, f_m$ as the file f_0 given by repeating:

- let $e_1, e_2, \ldots, e_m$ be the elements which we would be given by One Out on $f_1, f_2, \ldots, f_m$;
- let e_i be the smallest of these elements;
- remove e_i from f_i and add it to f_0.

Merging one file is just copying. The fact that merging two sorted files gives a sorted file and and the structure diagram

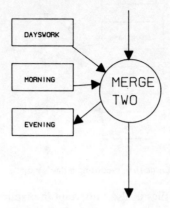

Small firm recording a day's work

show why two file merging is important in practice. A suitable generic procedure for merging two files is

```
procedure MergeTwo(a,b,c:STRING) is
          first,second:ELEMENT;
begin
          OneOut(a,first);OneOut(b,second);
          loop
                    if     Key(first)<Key(second)
                    then   OneIn (c,first);
                           exit when first = stop;
                           OneOut(a,first);
                    else   OneIn (c,second);
                           exit when second= stop;
                           OneOut(b,second);
                    end    if;
          end    loop;
end       MergeTwo;
```

Merging Two Files

The fact that merging any number of sorted files gives a sorted file and the structure diagram

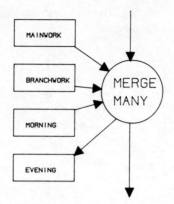

Large firm recording a day's work

show that merging many files is also important in practice, so you should welcome the generic procedure

```
procedure MergeMany(p:PROFILE;c:STRING) is
                i,least :INTEGER;
        begin
                Advance(p);least := 1;
                loop
                        for  i in  2..p.length
                        loop
                                if    Key(i) < Key(least)
                                then  least := i;
                                end   if;
                        end   loop;
                        OneIn(c,p.el(least));
                        exit when  p.el(least) = stop;
                        OneOut(p.name(least),p.el(least));
                end   loop;
        end       MergeMany;
```

Merging Many Files

Note that this procedure allows an arbitrary number of files to be merged because it uses the type PROFILE.

Before we can explain how merging is the key to all external sorting methods, we must introduce the concept of a run. A *run* is a sequence of elements a_i, $a_{i+1}, \ldots, a_j$ in a file such that

$a_i \leqslant a_{i+1} \leqslant \ldots \leqslant a_j$

a_i follows a larger element if it is not the first element of the file

a_j is followed by a smaller element if it is not the last element of the file

Runs in a file

A file is sorted if and only if it consists of precisely one run; every file is a sequence of runs. Suppose we have an element 'pause' which is less than the element 'stop' but greater than any other element. Let us say a file is *corrupt* if the last element of every run is a pause element. Any file can be corrupted by inserting a stop element after the last run, and a pause element after every other run.

Corrupted file

The idea behind all external sorting methods is the merging of runs, and corrupting files makes run-merging easier. All of our external sorting methods will use the generic task

```
task body RunAdministrator is --in body of package SortSearch
        first,second:ELEMENT;
begin
     loop
       select terminate;
         or accept Corrupt(a,c:STRING;runs:out INTEGER) do
                   Make(a);OneOut (a,first);runs:=1;
                   loop
                           OneIn(c,first);
                           exit when first= stop;
                           OneOut(a,second);
                           if    Key(first)>Key(second)
                           then  OneIn(c,pause);
                                 runs:=runs+1;
                           end   if;
                           first:=second;
                   end   loop;
               end;
         or accept TwoRunMerge(a,b,c:STRING;aend,bend:BOOLEAN) do
                   OneOut(a,first);OneOut(b,second);
                   loop
                           if   Key(first)<Key(second)
                           then OneIn(c,first);
                                aend:=(first=stop);
                                exit when Wait(c);
                                OneOut(a,first);
                           else OneIn(c,second);
                                bend:=(second=stop);
                                exit when Wait(c);
                                OneOut(b,second);
                           end if;
                   end   loop;
               end;
         or accept ManyRunMerge(p:PROFILE;c:STRING) do
                   Advance(p);
                   loop
                           least:=1;
                           for  i in 1..size
                           loop
                                if      p.el(i)<p.el(least)
                                then    least:= i;
                                end     if;
                           end loop;
                           OneIn (c,p.el(least));
                           exit when Wait(p.el(least));
                           OneOut(p.name(least),p.el(least));
                   end   loop;
               end;
```

```
or accept Kill(a,c:STRING) do
            Make(c);
            loop
                   ManyOut(a,v);
                   ManyIn (c,v);
                   exit when v(v'LAST)=stop;
            end    loop;
            Delete(c);
        end;
     end select;
   end  loop;
 end    RunAdministrator;
```

Run Administrator task body

The reason why external sorting methods work is that every call of Two Run Merge and Many Run Merge decreases the total number of runs, and one run files are sorted.

Our first external sorting method has the structure diagram

SIMPLE SORT IS

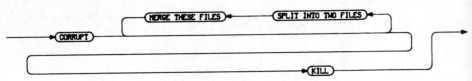

Simple Sort structure diagram

If we start with a file with 4 runs, the subalgorithm 'Split into two files' will give two files with 2 runs, and 'Merge these files' will give one file with 2 runs. The next application of 'Split into two files' will give two files with 1 run, the next application of 'Merge these files' will give 1 run, and this file is sorted. If we assume the subalgorithms, 'Split into two files' and 'Merge these files' are given by an environment Two File, we can convert our sorting algorithm into the generic procedure

```
procedure SimpleSort(from,to:STRING)is
          runs:INTEGER;
begin
          Make("A");Make("B");Make("C");
          Corrupt(from,"C",runs);
          loop
                 exit when runs=1;
                 TwoSplit("C","A","B");
                 TwoMerge("A","B","C");
          end    loop;
          Kill("C",to);Destroy("C");
          Destroy("A");Destroy("B");
end       SimpleSort;
```

Simple Sort procedure

Our second external sorting method avoids repeated file splitting by providing an extra file. The algorithm is given by

BALANCED SORT IS

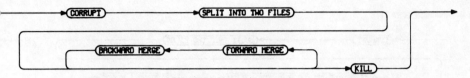

Balanced Sort structure diagram

If we start a file with 4 runs, the subalgorithm 'Split into two files' will give two files with 2 runs, the subalgorithm 'Forward Merge' will give two files with 1 run, and the subalgorithm 'Backward Merge' will give a file with 1 run and a file with 0 runs. If the environment Two File provides a suitable operation Balanced Merge, we can convert our sorting algorithm into a generic procedure.

```
procedure BalancedSort(from,to:STRING) is
          runs:INTEGER;
begin
          Make("A");Make("B");Make("C");Make("D");
          Corrupt(from,"C",runs);
          if     runs > 1
          then   TwoSplit("C","A","B");
                 loop
                     BalancedMerge("A","B","C","D",runs);
                     if   runs = 1
                     then Kill("C",to);exit;
                     end  if;
                     BalancedMerge("C","D","A","B",runs);
                     if   runs = 1
                     then Kill("A",to);exit;
                     end  if;
                 end  loop;
          else   Kill("C",to);
          end    if;
          Destroy("A");Destroy("B");Destroy("C");Destroy("D");
end       BalancedSort;
```

Balanced Sort procedure

The last of our external sorting methods avoids most file splitting without using an extra file. The idea is: as soon as a source file becomes empty, use it as the destination file. Suppose file C is the destination of a merge of a file A with 5 runs and a file B with 3 runs. After a while file A will have 2 runs, file B will have 0 runs and file C will have 3 runs. The first of our external searching methods would then copy the rest of A to C, but the new method uses file B as the destination of a merge of A and C. We can compare the two methods in the run tables

	A	B	C			A	B	C
Simple Search	5	3	0	Finonacci Search		5	3	0
	0	0	5			2	0	3
	3	2	0			0	2	1
	0	0	3			1	1	0
	2	1	0					
	0	0	2					
	1	1	0					
	0	0	1					

The algorithm for our new sorting method is

```
[BONACCI SORT IS
```

Fibonacci Sort structure diagram

If we have an environment Two File with an appropriate Fibonacci Merge, we can convert our algorithm into a generic procedure

```
procedure  FibonacciSort(from,to:STRING) is
           runs:INTEGER;source1,source2,sink:STRING(1..10);both:BOOLEAN
begin
           Make("A");Make("B");Make("C");
           Corrupt(from,"C",runs);
           source1:="A";source2:="B";sink:="C";both:=TRUE;
           loop
                   if     runs = 1
                   then   Kill(sink,to);exit;
                   else   TwoSplit (sink,source1,source2);
                          FibonacciMerge(source1,source2,sink,runs);
                   end    if;
           end    loop;
           Remove("A");Remove("B");Remove("C");
end        FibonacciSort;
```

Fibonacci Sort procedure

This procedure can be improved by not using Two Split when both source files become empty at the same time.

All of the external sorting methods we have described share the environment Two File, and they all have the same network diagram

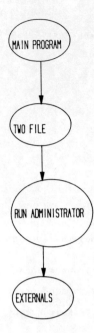

Network diagram for two file programs

You will not be surprised by the definition of the environment Two File:

```
task body TwoFile is
        s:STRING(1..80);e:ELEMENT;nomorex,nomorey:BOOLEAN;
begin
    loop
      select terminate;
        or accept TwoSplit(x,y,z:STRING) do
                  loop
                        s := y;
                        loop
                              OneOut(x,c);
                              OneIn (s,c);
                              exit when Wait(c);
                        end  loop;
                        if   c=stop
                        then exit;
                        elsif s = y
                        then s:= z;
                        else s:= y;
                        end  if;
                  end  loop;
              end;
        or accept SimpleMerge(x,y,z:STRING;runs:out INTEGER) do
                  runs:=0;
                  loop
                        TwoRunMerge(x,y,z,nomorex,nomorey);
                        runs:=runs+1;
                        exit when  nomorex and nomorey;
                  end  loop;
        or accept BalancedMerge(x,y,u,v:STRING;runs:out INTEGER) do
                  runs:=0;
                  loop
                        TwoRunMerge(x,y,u,nomorex,nomorey);
                        runs:=runs+1;
                        exit when nomorex and nomorey;
                        TwoRunMerge(x,y,v,nomorex,nomorey);
                        runs:=runs+1;
                        exit when nomorex and nomorey;
                  end  loop;
        or accept FibonacciMerge(x,y,z:STRING;runs:out INTEGER) do
                  runs:=0;both:=FALSE;
                  loop
                        TwoRunMerge(x,y,z,nomorex,nomorey);
                        runs:=runs+1;
                        exit  when nomorex and nomorey;
                        if    nomorex
                        then  s:=x;x:=z;z:=s;
                        elsif nomorey
                        then  s:=y;y:=z;z:=s;
                        end   if;
                  end  loop;
      end select;
    end  loop;
end  TwoFile;
```

Two File task body

We shall not resist the temptation to define an environment that allows an external sorting method to merge more than two files at a time.

```
task body     ManyFile is -- in body of SortSearch
              s:STRING(1..80);c:ELEMENT;i,k:INTEGER;
              function All(p:PROFILE) return BOOLEAN is
                      more :BOOLEAN;
              begin
                      k:=0;  more:= FALSE;
                      for i in 1..p.length
                      loop
                          if    p.el(i)=stop
                          then k:=i;
                          else more:=TRUE;
                          end if;
                      end loop;
                      return not more;
              end     All;
begin
      loop
        select terminate;
          or accept ManySplit(p:PROFILE;from:STRING) do
                      i:=1;
                      loop
                              s := p.names(i);
                              if    s/=from
                              then loop
                                          OneOut(from,c);
                                          OneIn (s    ,c);
                                          exit when Wait(c);
                                  end    loop;
                              end  if;
                              exit when c=stop;
                              if   i = p.length
                              then i := 1;
                              else i := i + 1;
                              end if;
                      end loop;
              end;
          or accept MergeSimple(p:PROFILE;to:STRING;
                                runs:out INTEGER) do
                      runs:=0;
                      loop
                      ManyRunMerge(p,to);runs:=runs+1;
                      exit when All(p);
                      end  loop;
              end  ;
          or accept MergeBalanced(p,q:PROFILE;
                                  runs:out INTEGER) do
                      i:=1;runs:=0;
                      loop
                              ManyRunMerge(p,q.names(i));
                              runs:=runs+1;
                              if    All(p) then exit;
                              elsif i=q.length
                              then  i:=1;
                              else  i:=i+1;
                              end   if;
                      end  loop;
              end;
          or accept  MergeFibonacci(p:PROFILE;to:STRING;
                                    runs:out INTEGER) do
                      runs:=0;
                      loop
                              ManyRunMerge(p,to);
                              runs := runs + 1;
                              if    All(p)
                              then  p.name(1):= to;exit;
                              elsif k/=0
                              then  p.name(k):= to;
                              end   if;
                      end  loop;
              end;
end    ManyFile;
```

Many File task body

206

This environment will be used in three programs for sorting a very large file of numbers. There will be one program for each of the external sorting methods we have described, but all of them have the network diagram

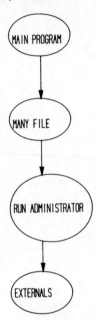

Network diagram for many file programs

The first program uses the Simple Sort algorithm

```
with      Generics
procedure  ManySort1 is
          function Wait(i:INTEGER) return BOOLEAN is
          begin return (i>1000000);end;
          function Key (i:INTEGER) return INTEGER is
          begin return (i);end;
          package SS is
                    new SortSearch(INTEGER,1000002,1000001,
                                              Wait,Key );
          p:SS.PROFILE(4);runs:INTEGER;
begin
          MakeP(p);Make("C");
          Corrupt("data","C",runs);
          while    runs>1
          loop
                    ManySplit(p,"C");
                    MergeSimple(p,"C",runs);
          end     loop;
          Kill   ("C","results");
          DestroyP(p);Destroy("C");
end      ManySort1;
```

ADA program: Many Sort 1

The second program uses the Balanced Sort algorithm

```
with      Generics
procedure  ManySort2 is
            function Wait(i:INTEGER) return BOOLEAN is
            begin return (i>1000000);end;
            function Key (i:INTEGER) return INTEGER is
            begin return (i);end;
            package SS is
                 new SortSearch(INTEGER,1000002,1000001,
                                           Wait,Key );
            p,q:SS.PROFILE(4);runs:INTEGER;
begin
        MakeP(p);MakeP(q);
        Corrupt("data",q.name(1),runs);
        if    runs>1
        then  ManySplit(p,q.name(1));
        loop
              MergeBalanced(p,q,runs);
              if   runs = 1
              then Kill(q.name(1),"results");exit;
              end  if;
              MergeBalanced(q,p,runs);
              if   runs = 1
              then Kill(p.name(1),"results");exit;
              end  if;
        end   loop;
        else  Kill(q.name(1),"results");
        end   if;
        DestroyP(p);Destroy(q);
end     ManySort2;
```

ADA program: Many Sort 2

The third program uses the Fibonacci Sort algorithm

```
with      Generics
procedure  ManySort3 is
            function Wait(i:INTEGER) return BOOLEAN is
            begin return (i>1000000);end;
            function Key (i:INTEGER) return INTEGER is
            begin return (i);end;
            package SS is
                 new SortSearch(INTEGER,1000002,1000001,
                                           Wait,Key );
            p:SS.PROFILE(4);runs,sink:INTEGER;manyempty:BOOLEAN;
begin
        MakeP(p);sink:=1;manyempty:=TRUE;
        Corrupt("data",p.name(1),runs);
        loop
              if    runs = 1
              then  Kill(p.name(sink),"results");exit;
              elsif manyempty
              then  ManySplit(p,p.name(sink));
              end   if;
              MergeFibonacci(p,sink,manyempty);
        end   loop;
        DestroyP(p);
end     ManySort3;
```

ADA program: Many Sort 3

Which of these three programs is the best? It is easy to see that the number of K-run-merges needed to sort a file with N runs is

$$(N - 1) / (K - 1)$$

If we want to sort a file with 1000 runs, this gives

- the first program needs 250 five-run-merges;
- the second program needs 500 three-run-merges;
- the third program needs 250 five-run-merges;
- any program using the environment Two File needs 1000 two-run-merges.

This comparison shows that multi-file sorting can be more efficient than two file sorting, and it indicates that the second program may be less efficient than the first and third programs.

We can turn any external sorting method into a mixed sorting method by using an internal sorting method to eliminate all short runs. Suppose a rendezvous on the entry Corrupt in the task Many File uses an internal sorting method to change our file with 1001 runs to a file with 101 runs. This change would reduce the number of run merges needed by our three programs for sorting the file to a tenth— a dramatic increase in efficiency. Another way of making external sorting methods more efficient is to use the Many_In, Many—Out operations of File_Administrator intead of the One_In, One_Out operations used by the version of Run_Administrator that we have given. If the underlying computer allows data transmission at the same time as computation, we can write File_Administrator and Run-Administrator in such a way that they run concurrently. The possibilities are legion; the best sorting method for one kind of data and computer configuration may not be the best sorting method for another kind of data or another computer configuration. The whole topic of sorting is a good illustration of the problem solving maxim: once you have a solution to a problem, you can almost always find a more efficient solution.

Exercise

The body of our environments Private Data Base and Public Data Base uses the environment Table Manipulator but we have not yet defined the body of Table Manipulator. Your exercise is to define two versions of this body—a version with 'type TABLE is array (1 . . size) of STRING (1 . . 10)' and a version with 'type TABLE is new INOUT_FILE'. Whenever possible use the generic procedures from this chapter and ideas from the tree version of this body:

```
package TableManipulator is
        type ACTION is (proc  :,sum,differ,join,project
                       ,same  efore,after,insert,joining,looking);
        type TABLE   is priva :;
        function Mix(u,v:TAB E;a:ACTION;match:BOOLEAN)
                                        return TABLE;
        function Modify(u:TABLE;l:LINE;a:ACTION)
                                        return TABLE;
        function MatchOnFirstName(l,u:LINE)  return LINE;
        function Permute          (l,u:LINE)  return LINE;
        function Convert          (l  :LINE)  return TABLE;
        task     TableAdministrator is
                 entry  GetTable(t:out TABLE);
                 entry  PutTable(t:in  TABLE);
        end      TableAdministrator;
        bad_parameters:EXCEPTION;
private type ELEMENT;
        type TABLE is access ELEMENT;
        type ELEMENT(size:INTEGER) is record
                            value:array(1..size) of STRING(1..10);
                            left,right:TABLE;
                            end    record;
end     TableManipulator;


package body TableManipulator is
            i,j,k,comparison,result:TABLE;found:BOOLEAN;
        function  Mix(u,v:TABLE;a:ACTION;match:BOOLEAN)
                                        return TABLE      is
        begin
                result:= null;
                if    not match then raise bad_parameters;
                elsif a=sum     then Traverse(v,sum);
                -- Traverse is a procedure given below
                -- Traverse(v,sum) copies v to result
                else   comparison := v;
                end    if;
                Traverse(u,a);return result;
        end     Mix;

        function  Modify(u:TABLE;l:LINE;a:ACTION)
                                return  TABLE     is
        begin
                result :=null;
                k:= Convert(l);
                Traverse(u,a);return result;
        end     Modify;
-- the body of the task TableAdministrator and the functions
-- Convert,Permute,MatchOnFirstName,Extension,Rearrange
-- are left to the reader;

        procedure Traverse(t: in out TABLE;a:ACTION) is
        begin
            if   t = null
            then if   a=insert
                 then t:= new ELEMENT
        (size=>i.size, value=>i.value,left=>null,right=>null);
                 end  if;
                 return;
            else case a is
                 when product =>i:=t;found:=FALSE;
                                Traverse(comparison,looking);
                                if    found
                                then Traverse(result,insert);
                                end  if;
```

```
                when sum     =>i:=t;Traverse(result,insert);
                when differ  =>i:=t;found:=FALSE;
                             Traverse(comparison,looking);
                             if   not found
                             then Traverse(result,insert);
                             end  if;
                when project =>i:=Rearrange(t);
                             Traverse(result,insert);
                when join    =>j:=t;Traverse(comparison,joining);
-- discover how the remaining cases exploit "TABLES as trees"
                when joining =>if     j.value(1) > t.value(1)
                             then  Traverse(t.right,a);return;
                             elsif j.value(1) < t.value(1)
                             then  Traverse(t.left ,a);return;
                             else  i:=Extension(t,j);
                                   Traverse(result,insert);
                             end   if;
                when same    =>if     k.value(1) > t.value(1)
                             then  Traverse(t.right,a);return;
                             elsif k.value(1) < t.value(1)
                             then  Traverse(t.left ,a);return;
                             else  i:=t;Traverse(result,insert);
                             end   if;
                when above   =>if     k.value(1) < t.value(1)
                             then  i:=t;Traverse(result,insert);
                             else  Traverse(t.right,a);return;
                             end   if;
                when below   =>if     k.value(1) > t.value(1)
                             then  i:=t;Traverse(result,insert);
                             else  Traverse(t.left ,a);return;
                             end   if;
                when others  =>if     i.value > t.value
                             then  Traverse(t.right,a);return;
                             elsif i.value < t.value
                             then  Traverse(t.left ,a);return;
                             else  found:=TRUE;        return;
                             end   if;
                end   case;
                Traverse(t.right,a);Traverse(t.left,a);
           end  if;
      end    Traverse;
end TableManipulator;
```

Table Manipulator package

Chapter 8

Computers and People

'The reasonable man adapts himself to the world:
the unreasonable one persists in trying to adapt the
world to himself. Therefore all progress depends on
the unreasonable man.'

G. Bernard Shaw, *Man and Superman.*

8.1 Prehistory

The history of computing has been made by people with a vision of what could be
done with information once it has been recorded. Our account will focus on these
people and their vision, but we will not forget the dangers of the computer revolution. Perhaps we should start as we shall finish with the dangers. The scene is the
Inca community in the Andes Mountains many centuries ago. The ruler of the
community based his decisions on very detailed information about his people. All
important facts were encoded on knotted cords called quipus and this quipus were
carried by runners to the capital Cuzco. A typical quipu might record the census
information for a village—the knots on one cord telling how many inhabitants were
over sixty, the knots on another saying how many were under ten. When the Inca
society was conquered by the Spaniards in the sixteenth century the last ruler
Atahuallpa sent orders by quipus from his prison. His subjects duly collected the
gold for his ransom, but the Spaniards killed him even so. The quipu system con-
tinued; it became a useful tool to enslave the people. Just as the quipu system con-
solidated the power of both the Inca rulers and the Spanish conquerors, computers
consolidate the power of modern quipu-receivers; data base sytems and governments both good and bad. As McLuhan says 'The computer is by all odds the most
extraordinary of all technological clothing ever devised by

211

212

Typical Inca Quipu

man, since it is the extension of our central nervous system. Beside it the wheel is a mere hula-hoop, though that is not to be despised entirely.' The Inca civilization did not have the wheel, but it lasted for several centuries.

'The vision that pervades the history of computing can be expressed as: Calculation and reasoning, like weaving and ploughing, are work not for human souls but for clever combinations of iron and wood. If we spend our time in doing what a machine could do faster than ourselves, it should only be for exercise as we swing dumbbells or for amusement as we dig our garden, but not in any hope of so working our way to the truth.'

The ADA type Boolean is not named after the author of this quotation, Mary Boole, but her logician husband. Her vision is similar to that of Ramon Lull in the thirteenth century. Lull so dazzled his contemporaries that he was known as Doctor Illuminatus. In the hope of freeing philosophy from theology, of basing Reason on Doubt not Faith, he invented a device called Ars Magna. It consisted of a number of circles, which could be set in various positions. By setting the circles appropriately one could find the answer to one's questions. The first useful computing devices tackled less important calculations than Lull's Ars Magna.

The most successful early calculator was designed by Blaise Pascal in 1642. Pascal was brought up by his taxcollecting father, and his calculator may have been inspired by seeing his father laboriously summing long lists of numbers.

Pascal was granted a royal monopoly for making his machines, and from time to time he demonstrated them to fashionable audiences.

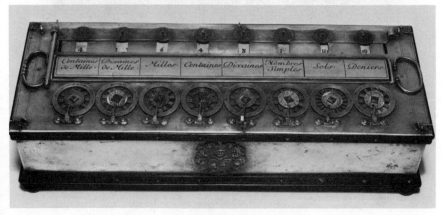

Pascal and his machine

(Reproduced by permission of IBM)

However computing was only one of Pascal's interests, he more or less founded probability theory, he made important contributions to geometry and physics, he suffered severe physical illness and a mystical religious experience, he wrote a masterpiece—his *'Pensées'*.

Our next computer designer is Gottfried Leibnitz. His machine could not only add and subtract like Pascal's, it could also multiply and divide. It is a sign of the times that Leibnitz sent one of his machines to the Tsar of Russia, just as Pascal presented one of his to Queen Christina of Sweden.

Leibnitz and his machine

(Reproduced by permission of IBM)

Leibnitz is best known for his mathematics—many feel that he invented the calculus—but he was also a great philosopher. The driving force behind his philosophy was his belief that all disputes, however complex, could be settled by calculation if only we had a characteristica universalis, a general way of representing information. He invented one of the first formal logic systems, and he tried to solve diplomatic problems by formal symbol manipulation. When there were four candidates for the throne of Poland, he tried to find the most suitable candidate by formal manipulation of sixty logical propositions. Because of his formal arguments, showing that Louis XIV should satisfy his colonial ambitions by invading Egypt instead of marching into the Rhine provinces, he was sent to Paris. Leibnitz was intimately involved in practical affairs, but his 'charactiristica universalis' is a vision no different from Lull's Ars Magna.

8.2 Babbage and Countess Ada Lovelace

We have defined an algorithm as a complete, unambiguous procedure for solving a problem in a finite number of steps. The word algorithm is a corruption of the name of the ninth century Arab mathematician al-khowarizi who gave step-by-step instructions for solving problems in his book Kitab al jabr w al-mugabala. The conceptual leap from the calculators of Leibnitz and Pascal to the computers of today could not have been made until some visionary thought of a way to represent algorithms, to incorporate algorithms in a machine. As Charles Babbage (1791-1871) was that visionary, we shall describe the way he represented algorithms in his Analytic Machine in some detail. But first the man himself in two quotes from his biographer Mabott Moseby:

'He made acute analyses of the pin-making industry and the printing trade. He analyzed the economics of the Post Office as a result of which Sir Rowland Hill introduced the penny post. He studied insurance records and published the first comprehensive treatise on actuarial theory and the first reliable life tables. He invented the bibliograph and the opthalmoscope . . . He took a life-long interest in ciphers and deciphering, and he could pick any lock. He wrote a ballet and invented coloured lighting for the stage.'

'Contentious, contrary, sarcastic and cynical, there was another side to his character. The son of a rich London banker, he was no unapproachable recluse hiding away in an ivory tower. He was a sociable, witty and loveable man, with a ringing, hearty laugh and an unfailing fund of amusing anecdotes on which he constantly dined out. Invitations to his Saturday-evening parties were eagerly sought, the indispensable qualifications being intellect, beauty or rank. His son records that some 200 to 300 people would attend one of his gatherings, and that on one occasion no fewer than seven bishops were present.'

Babbage was an eccentric genius and often made bitter and somewhat unjust attacks on those who did not appreciate his talents. Not that he went without recognition in his time. The British Government gave him 17000 pounds for the development of his Difference and Analytic engines; the Lucasian professor of mathematics in Cambridge resigned his position in favour of the 27 year old Babbage. The professorship is that of Newton, and the duties are very light: to live within three miles of Great St. Mary church during the university term.

III

GENERAL PLAN OF ENGINE No. 1

Plan 25, dated August 6, 1840

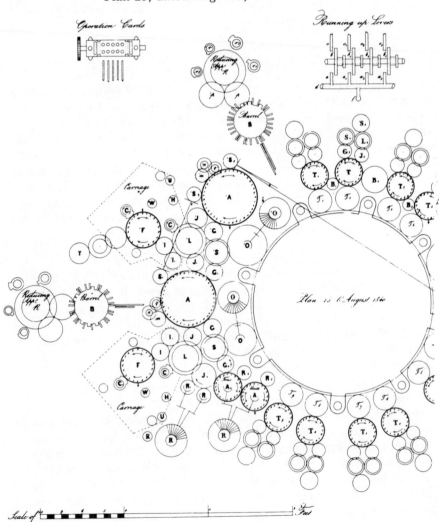

Although he was professor for eleven years, there is no record of Babbage ever giving a lecture.

We shall not describe the Difference engine here because it closely resembles earlier calculating devices: the values of variables correspond to the positioning of gear wheels and the user has to communicate the steps

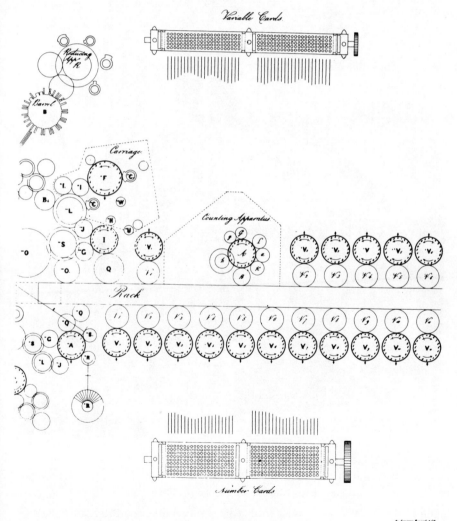

Babbage's Analytic Engine

(Reproduced by permission of Dover Publications, Inc.)

218

in an algorithm one by one, she has to decide on the next operation to be performed by the device. We shall describe Babbage's design for his Analytic Engine because it was closer to modern computer designs than anything else before 1940 in that it incorporated a way of representing a complete algorithm. Algorithms were to be incorporated in the analytic engine as a sequence of cards, just as the patterns woven by Jacquard's automatic looms were determined by sequences of cards. With this in

210. Métier Jacquard.

Jacquard's Loom

(Reproduced by permission of IBM)

mind, read what Babbage himself wrote (the words in the parentheses are mine):

'The Jacquard loom (computer) is capable of weaving any design (solving any problem) which the imagination of man may conceive. It is also constant practice for skilled artists (programmers) to be employed by manufacturers in designing patterns (writing programs). These patterns are then sent to a peculiar artist (typist) who, by means of a certain machine (a terminal) punches holes in a set of pasteboard cards in such a manner that when those cards are placed in a Jacquard loom (submits the program to the computer), it will then weave upon its produce the exact pattern designed by the artist (the computer executes the program). Now the manufacturer may use, for the warp and weft of his work, threads which are all of the same colour; let us suppose them to be unbleached or white threads (the computer can execute the program without input). In this case the cloth will be woven all of one colour; but there will be a damask pattern upon it such as the artist designed (the computer can check if a program is syntactically correct). But the manufacturer might use the same cards and put into the warp threads of any other colour (a program can be executed with various inputs). Every thread might even be of a different colour or of a different shade of colour; but in all cases the *form* of the pattern will be precisely the same—the colours only will differ. The analogy of the Analytic Engine with this well-known process is nearly perfect.'

No wonder Babbage acquired a self-portrait of Jacquard woven by one of his looms (controlled by 24000 cards each with room for 1050 punch holes); but Babbage also realised that Jacquard's card-reading mechanism would limit his analytic engine to straight line algorithms, and he devised a mechanism for repeating and skipping cards that reminds one of ADA's *loop* and *exit* constructs.

The programming language ADA gets its name from Countess Ada Lovelace who thought hard about how the Analytic Engine should be programmed. She was the daughter of the poet Byron, and they are buried in the same grave. She was also a friend of Babbage and appreciated his ideas. Perhaps too much so; it seems she had to pawn some of her jewels to pay for the losses Babbage's system generated, when he, she and her husband went horse-racing. Babbage is said to have been a bad lecturer, but Count Mandelbrae (later a Garibaldi general, and prime minister of Italy) understood his lectures in Turin sufficiently well to take notes which Ada Lovelace transformed into the first readable introduction to the art of programming. We can sum up this section by quoting her:

'The analytic engine weaves algebraical patterns, just as the Jacquard loom weaves flowers and leaves.'

To appreciate this remark you should know that our word algrebra is derived from the title of the book by al-Khowarizi in which algorithms first appear.

8.3 Turing and Von Neumann

Although the designers of the first working computers in the 1940's had to rediscover Babbage's ideas, at least two of them deserve special mention because their

220

own ideas have been very influential. Our first modern designer, A. M. Turing (1912-1954), made important contributions to mathematical logic and biology. In 1936 he published a careful analysis of what could and what could not be done by computation. He introduced a precise mathematical definition of a program, he advanced the thesis that

'there is an algorithm to solve a problem if and only if there is a program that solves the problem

and he described a problem that could not be solved by an algorithm. Somewhat earlier another logician, A. Church, had presented a related thesis and unsolvable problem, but he was not a computer designer. As every programmer should know that there are problems which cannot be solved by a computer, we shall describe Turing's Halting Problem:

To devise a program ACCOUNTANT such that
1) executing ACCOUNTANT with a program P and data D as input gives output 'P on D stops' if execution of P does in fact stop;
2) executing ACCOUNTANT with a program P and data D as input gives output 'P on D does not stop' if the execution of P with input D does in fact continue forever;
3) the execution of ACCOUNTANT with a program P and data D as input always stops.

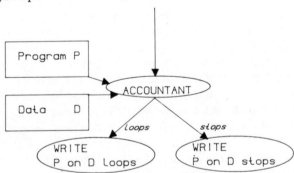

ACCOUNTANT structure diagram

Programmers, being human, often produce programs that never stop, that get into loops they cannot get out of. One could avoid wasting computer time on such programs if one had a program ACCOUNTANT; before running a program P on data D, the computer could input P and D to ACCOUNTANT and refuse to run P if ACCOUNTANT gave 'P on D does not stop'. Unfortunately we cannot have a program ACCOUNTANT, because if we could the following program could be solved.

To devise a program INVERT such that
1') execution of INVERT with a program P as input will not stop if execution of P with P as input stops;

2') execution of INVERT with a program P as input stops if execution of P with P as input does not stop.

The structure diagram for INVERT could be

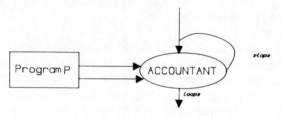

INVERT structure diagram

But this cannot be; INVERT cannot exist, because we could give it itself as input and get the absurdity

1") execution of INVERT with INVERT as input will not stop, if execution of INVERT with INVERT as input stops;

2") execution of INVERT with INVERT as input stops, if execution of INVERT with INVERT as input does not stop.

Turing made this argument precise by first distinguishing between the execution of a program P and its representation, then solving the program

To devise a program UNIVERSAL such that executing UNIVERSAL with program P and data D as input gives the same output as executing P with D as input

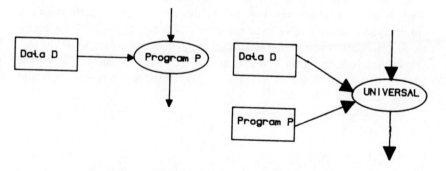

UNIVERSAL structure diagram

There are many ways of solving this problem; every translator for a programming language, indeed every modern computer, is a realization of UNIVERSAL.

In 1950 Turing discussed the question of whether machines can 'think' or not. He presented various arguments for believing that machines cannot think:

(Theological)

'Thinking is a function of man's immortal soul.'

222

(Heads in the sand)
'The consequences of machines thinking would be too dreadful. Let us hope and believe that they cannot do so.'
(Mathematical)
'The Halting problem is unsolvable.'
(Consciousness)
'Not until a machine can write a sonnet or compose a concerto because of thoughts and emotions felt, and not by chance full of symbols, could we agree that machine equals brain—that is not only write it but know that it had written it. No mechanism could feel (and not merely artificially signal, an easy contrivance) pleasure at its successes, grief when its valves fuse, be warmed by flattery, be made miserable by its mistakes, by charmed by sex, be angry or depressed when it cannot get what it wants.'— Jefferson's Lister Oration.
(Originality)
'The Analytical Engine has no pretensions to originate anything. It can do whatever we know how to order it to perform.'—Countess Ada Lovelace.
(False Induction)
'I grant you that you can make machines do all the things you have mentioned, but you will never be able to make one to do X.'

Turing accepted none of these arguments, he gave a precise meaning to the question 'Can Machines think?', and he believed that the only reasonable answer to the question was 'Yes'. He writes:

'The imitation game is played with three people, a man (A), a woman (B), and an interrogator (C) who may be of either sex. The interrogator stays in a room apart from the other two. The object of the game for the interrogator is to determine which of the other two is the man and which is the woman. He knows them by labels X and Y, and at the end of the game he says "X is A and Y is B" or "X is B and Y is A". The interrogator is allowed to put questions to A and B thus:

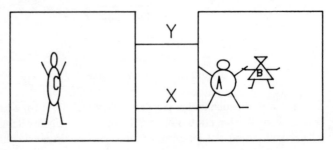

Turing's Test

C: "Will X please tell me the length of his or her hair". Now suppose X is actually A, then A must answer. It is A's object in the game to try and cause

C to make the wrong identification. His answer might therefore be: "My hair is shingled, and the longest strands are about nine inches long."

The object of the game for the third player (B) is to help the interrogator. The best strategy for her is probably to give truthful answers. She can add such things as:

"I am the woman, don't listen to him!"

to her answers, but it will avail nothing as the man can make similar remarks.

We now ask the question "What will happen when a machine takes the part of A in this game?" Will the interrogator decide wrongly as often when the game is played like this as he does when the game is played between a man and a woman? These questions replace our original "Can machines think?".'

Several cases of people conversing for several minutes with a computer program in the belief that they were communicating with a person have been reported. Another example of the computer's success in a variant of the Imitation game was reported in the New York Times, 12 March 1965:

'A computerized typewriter has been credited with remarkable success at a hospital here in radically improving the condition of several children suffering an extremely severe form of childhood schizophrenia. What has particularly amazed the number of psychiatrists is that the children's improvement occurred without psychotherapy; only the machine was involved. It is almost as much human as it is machine. It talks, it listens, it responds to being touched; it makes pictures or charts, it comments and explains, it gives information and can be set up to do this in any order . . . the machine was able to bring the autistic children to respond because it eliminated humans as communication factors. Once the children were able to communicate, something seemed to unlock in their minds, apparently enabling them to carry out further normal mental activities that had eluded them earlier.'

The last visionary in our history of computers is van Neumann (1903–1957), a mathematician whose contributions to quantum physics and economics were just as important as his contributions to logic, algebra and other areas of pure mathematics. At Princeton in 1945 von Neumann held a series of lectures on the design of computers in which he stressed the important of storing programs as well as data values. The advantage of this is that the computer can go from one step in an algorithm to the next at electronic speeds, not mechanical—the advantage of travelling by flying, not walking. Running a program on a von Neumann machine consists of repeating a FETCH-EXECUTE cycle

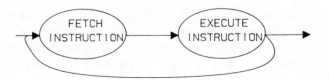

Fetch—Execute structure diagram

A typical instruction consists of an operation code and addresses for one or more arguments and results. When the computer fetches such an instruction from the store, it separates the operation code and the addresses. The structure diagram for this is

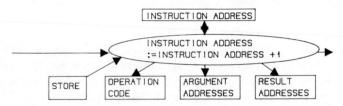

Fetch structure diagram

When the computer executes an instruction, it begins by getting the arguments from the store, it obeys the operation, then it puts the results into the store. If we are to avoid the straight jacket of straight line programs, we must allow some instructions to change the value of Instruction_Address. The appropriate structure diagram is

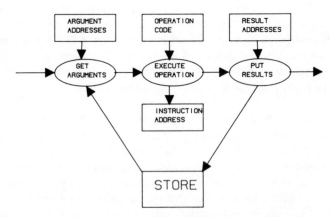

Execute structure diagram

In a long report on 'planning and coding problems for an electronic computing instrument' written by von Neumann and his collaborator Goldstine we find the first description of structure diagrams like those we have been using in this book: (an abbreviated quote)

'The actual code for a problem is that sequence of coded symbols that has to be placed in the memory in order to cause the machine to perform the desired and planned sequence of operations, which amount to solving the problem in question . . . In planning a coded sequence the thing one should keep primarily in mind is its functioning while the process that it controls goes through its course. We therefore propose to begin the planning by laying out the flow diagram

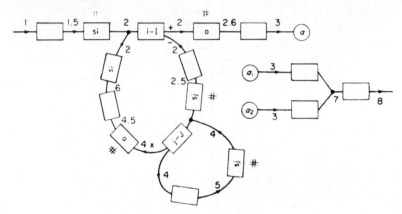

Von Neumann's flow diagrams

(Reproduced by permission of Pergamon Press)

Von Neumann was fascinated by the differences between the human brain and his design for a computer in the Princeton lectures. In his computer design there is just one computation and any error is fatal, whereas there can be many simultaneous computations in the brain and it can recover from errors by comparing the results of several computations. He described how one could build a reliable computer from unreliable components by duplicating computations. Some modern computers are based on another of his biologically motivated computer designs—a highly parallel *cellular automaton.*

A cellular automaton is a computer with just one instruction but a very special kind of store. The store of a Princeton Machine is finite and individual components are distinguished by addresses, whereas the store of a cellular automaton is infinite, and individual components are distinguished by a number and a letter, just as the individual squares on a chess board are distinguished by giving a row and a column.

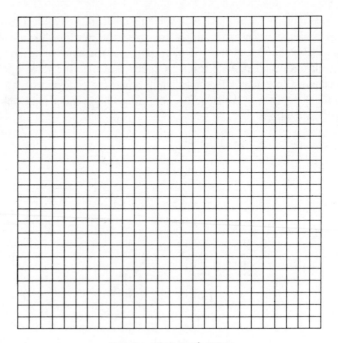

Cellular automaton's store

The one instruction of the cellular automaton determines the new value of *every* cell from the old values of the cell and its neighbours.

One fascinating cellular automaton that has been simulated at many computer installations is Conway:s 'LIFE'. In LIFE the values of the cell are ALIVE or DEAD, every cell starts with one of these values, and only finitely many cells start with the value ALIVE. The instruction is defined as follows:

Let n be the number of neighbours of cell (i,j) that are ALIVE;
(QUIESCENT)   If cell (i,j) is DEAD and $n \neq 3$, it remains DEAD;
(BIRTH)   If cell (i,j) is DEAD and $n = 3$, it becomes ALIVE;
(LONELY)   If cell (i,j) is ALIVE and $n \leqslant 1$, it becomes DEAD;
(SURVIVAL)   If cell (i,j) is ALIVE and $1 < n < 4$, it remains ALIVE;
(OVERCROWDED) If cell (i,j) is ALIVE and $n \geqslant 4$, it becomes DEAD.

There may be no bound on the number of ALIVE cells, but QUIESCENT ensures that this number is always finite.

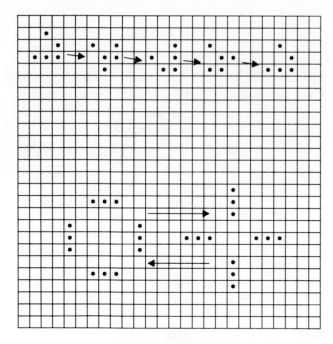

Glider and Traffic Lights

Another fascinating cellular automaton was designed by von Neumann to illustrate self-reproduction. It has close analogues of the genes and DNA-splitting characteristic of biological reproduction.

Most existing computers are elaborations of the design presented by von Neumann in his Princeton lectures. By now there are about a million computers at work around the world; the exact figure depending on where one draws the line between computers and calculators. We should think what these computers are doing, not how many there are.

8.4 Epilogue

We have kept a very important theme for our closing paragraphs in the hope that it will be remembered when most of our other themes have been forgotten. The programs the reader and her colleagues develop may affect the lives of other people. Imagine you are a member of parliament in the nineteenth century, attending a debate on the transport situation.

Social questions

Imagine the arrival of an angel who says:

Although you do have trains and canal boats you mostly use horses for transporting goods and people. Horses are slow, eat much hay, and need looking after. I can give you a device which travels fast, devours what springs up from the ground and requires little attention. It will ease your transport problem but the human price is high—five thousand will be killed each year and far more maimed. How do you reply?'

If you would say NO, would you be thinking of the future dead and wounded or the manufactures of horse carriages?

Some people benefit from the existence of a computer program, and some people are harmed. Computer programs, like those in this book, are being used to produce newspapers. One social consequence is that many typographers are out of work, another is that there are more misprints and the quality of the newspapers has deteriorated. Who benefits when computers replace typographers—perhaps the advertisers pay less, perhaps the newspaper owners make larger profits. A new computer program raises the same social problems as any other technological innovation. The social problems caused by the fact that there were many thousands of Jacquard looms forty years after their invention are very similar to those due to the computers we have around us today. Industrialization itself has been accused of turning happy craftsmen into unhappy factory workers, and providing cheap, shoddy substitutes for quality goods.

Because information gives power to decision makers new computer programs also raise social problems, that differ from those caused by other technological innovations. Consider the care of a doctor for his patients. One form of organisation is for each doctor to have his own patients who he knows personally. The disadvantage of such a decentralised organisation is that each doctor has to be prepared for night calls every night. A form of organisation, that does not have this disadvantage, is for a group of doctors to share patients, to form a medical centre and share patient journals. As soon as these patient journals are kept in a computer, the medical centre will tend to grow larger and acquire administrators. There are two disadvantages of such a computerised, centralised organisation—alienation and privacy. Alienation is felt by the patients whose lives are affected by decisions made by administrators and doctors who do not know them personally. Many patients will also be worried by the fact that many people will have access to private information in their medical journals. How can such patients be sure that the curious and the criminal do not have access to this private information also? The quality of a data base system depends crucially on its solution of the privacy problem. Programmers should not concentrate entirely on devising correct and efficient programs, they should also strive to improve their quality. Programmers should think about the social consequences of the programs they write, before they start writing them.

Exercise

Officials in the European Common Market have proposed a common passport for citizens of the market countries. In this passport there would be a page that could be read by a terminal at each frontier so the police could keep track of everyone's travelling. Naturally this proposal has been opposed by many human rights groups. Write an article for a right wing newspaper about how the proposed passport would help the police catch terrorists and/or write an article for a left wing newspaper about how the proposal would make opposition to the official government policies more perilous.

Program and Package Index

Main Index

233